# SpringerBriefs in Applied Sciences and Technology

## Computational Intelligence

**Series editor**

Janusz Kacprzyk, Warsaw, Poland

**About this Series**

The series "Studies in Computational Intelligence" (SCI) publishes new developments and advances in the various areas of computational intelligence—quickly and with a high quality. The intent is to cover the theory, applications, and design methods of computational intelligence, as embedded in the fields of engineering, computer science, physics and life sciences, as well as the methodologies behind them. The series contains monographs, lecture notes and edited volumes in computational intelligence spanning the areas of neural networks, connectionist systems, genetic algorithms, evolutionary computation, artificial intelligence, cellular automata, self-organizing systems, soft computing, fuzzy systems, and hybrid intelligent systems. Of particular value to both the contributors and the readership are the short publication timeframe and the world-wide distribution, which enable both wide and rapid dissemination of research output.

More information about this series at http://www.springer.com/series/10618

Leticia Cervantes · Oscar Castillo

# Hierarchical Type-2 Fuzzy Aggregation of Fuzzy Controllers

 Springer

Leticia Cervantes
Division of Graduate Studies
Tijuana Institute of Technology
Tijuana, Baja California
Mexico

Oscar Castillo
Division of Graduate Studies
Tijuana Institute of Technology
Tijuana, Baja California
Mexico

ISSN 2191-530X          ISSN 2191-5318   (electronic)
SpringerBriefs in Applied Sciences and Technology
ISBN 978-3-319-26670-1          ISBN 978-3-319-26671-8   (eBook)
DOI 10.1007/978-3-319-26671-8

Library of Congress Control Number: 2015954984

Springer Cham Heidelberg New York Dordrecht London

Printed on acid-free paper

Springer International Publishing AG Switzerland is part of Springer Science+Business Media
(www.springer.com)

# Preface

This book focuses on the fields of fuzzy logic, granular computing and also considering the control area. These areas can work together to solve various control problems, and the idea is that this combination of areas would enable even more complex problem solving and better results. In this book, we test the proposed method using two benchmark control problems: the total flight control and the problem of water-level control for a three-tank system. When fuzzy logic is used to make it easy to perform the simulations, these fuzzy systems help to model the behavior of a real systems; using the fuzzy systems, fuzzy rules are generated, and with this, they can generate the behavior of any variable depending on the inputs and linguistic value. For this reason, we consider in this work the proposed architecture using a combination of simple individual fuzzy systems and with this improve the behavior of the complex control problems.

This book is intended to be a reference for scientists and engineers interested in applying fuzzy logic techniques for solving problems in intelligent control. This book can also be used as a reference for graduate courses such as soft computing, intelligent control, fuzzy control. We consider that this book can also be used to get novel ideas for new lines of research or to continue the lines of research proposed by the authors of the book.

In Chap. 1, we begin by offering a brief introduction of the potential use of the hierarchical control method in different real-world applications. We describe the use of this method using type-2 fuzzy systems for aggregation of results in problems of intelligent control of nonlinear plants. We also mention other possible applications of the proposed control approach.

We describe in Chap. 2 the basic concepts, notation, and theory of fuzzy logic, fuzzy control, granular computing, genetic algorithms, and aviation. This chapter overviews the background, main definitions, and basic concepts, useful for the development of this research work.

We describe in Chap. 3 the proposed method for hierarchical modular control of complex plants and then the particular control problems that are used to test the proposed method are explained. A new hierarchical approach combining multiple controllers is used as a new method for global control. This is critical for complex control problems that can be solved by dividing it into several simple controllers.

Chapter 4 is devoted to show the simulation results of the proposed method for control when applied to the particular nonlinear plants that are considered for validating the proposed approach. In particular, the three-tank problem and an aircraft dynamic system are considered. The simulation results are compared against simple fuzzy control. In addition, genetic algorithms are used to optimize the fuzzy controller and achieve a fair comparison. Statistical tests are used to show the advantage of the proposed method for control.

We offer in Chap. 5 the conclusions of this work, as well as some future research work is envisioned. Basically, a new control method was proposed, and then different cases of control were studied. The first case was the water control using three tanks and five valves connected between the tanks. To know whether the proposed method could give good results, first the control problems were studied in more detail. These details were obtained working first with individual controllers using type-1 and type-2 fuzzy systems and their optimization with genetic algorithms. Then, the proposed method was applied using type-1 and type-2 fuzzy systems for the problems of the three tanks and the aircraft system.

We end this preface of the book by giving thanks to all the people who have helped or encouraged us during the writing of this book. First of all, we would like to thank our colleague and friend Prof. Janusz Kacprzyk for always supporting our work and for motivating us to write our research work. We would also like to thank our colleagues working in Soft Computing, which are too many to mention each by their name. Of course, we need to thank our supporting agencies, CONACYT and DGEST, in our country for their help during this project. We have to thank our institution, Tijuana Institute of Technology, for always supporting our projects. Finally, we thank our respective families for their continuous support during the time that we spend in this project.

Leticia Cervantes<br>
Oscar Castillo

# Contents

# Chapter 1
# Introduction

There are many real world complex control problems on which traditional approaches produce limited results. For this reason, decomposing a complex problem into many simple control problems can be a good idea. In this way each of the simple problems can be solved by a simpler controller and then the overall control could be obtained by an appropriate combination of the outputs of the individual controllers.

This book focuses on the fields of fuzzy logic, granular computing and also considering the control area. These areas can work together to solve various control problems, the idea is that this combination of areas would enable even more complex problem solving and better results. As we mentioned before this document explain a proposed method were some areas are involved: fuzzy logic and optimization, control area and granularity.

There are already several studies that have proposed control techniques [1], which use more than one controller, for example in [2] the authors design a fuzzy cascade controller scheme, which consists of two fuzzy controllers arranged on a cascaded topology. The parameters of the controller are optimized by means of a hierarchical genetic algorithm. The fuzzy cascade scheme comprises two controllers located in two loops. Another approach proposed in [3] uses a fuzzy rule-based combination of linear and switching state-feedback controllers, for nonlinear systems subject to parameter uncertainties. The switching state-feedback controller is applied to drive the system states toward the origin [4].

In this book we test the proposed method using 2 benchmark problems: the total flight control and the problem of water level control for a 3 tank system. When fuzzy logic is used it make it easy to performed the simulations, these fuzzy systems help to model the behavior of a real systems, using the fuzzy systems fuzzy rules are generated and with this can generate the behavior of any variable depending on the inputs and linguistic value. For this reason we consider in this work the proposed architecture using fuzzy systems and with this improve the behavior of the complex control problems.

The architecture was created to reduce the complexity of a control system and work with this in a simple way. Fuzzy controllers were designed with the objective to control the open and closeness of the valves in the first case of study and in the second case to maintain the stability of an airplane by controlling elevators, ailerons

© The Author(s) 2016
L. Cervantes and O. Castillo, *Hierarchical Type-2 Fuzzy Aggregation of Fuzzy Controllers*, SpringerBriefs in Computational Intelligence, DOI 10.1007/978-3-319-26671-8_1

and rudder. Also a genetic algorithm is used to optimize the fuzzy systems of the proposed method. The first case of study was considered because some works used simple water tanks and we want to add complexity to this problem and test the method with a more difficult case, the second case of study as selected because in the total control of the airplane we can observe the higher complexity of the problem and was selected as second case of study but with more complexity (using the total control and noise).

This book is organized as follows: In Chap. 2 Theory and Background is presented, Chap. 3 Control Problem and Proposed Method is explained, in Chap. 4 Simulation Results are illustrated, Conclusions and Future work is presented in Chap. 5, and finally References and Appendix are present.

# References

1. Attia, A.-F.: Hierarchical fuzzy controllers for an astronomical telescope tracking. Appl. Soft Comput. **9**(1), 135–141 (2009)
2. No, T.S., Mina, B.M., Stone, R.H., Wong, K.C.: Control and Simulation of Arbitrary Flight Trajectory-Tracking. Department of Aerospace Engineering, Chonbuk National University, Deokjin Dong, Chonju NSW 2006, pp. 560–756
3. Lam, H., Leung, F.: Stability analysis of discrete-time fuzzy-model-based control systems with time delay: time delay-independent approach. Fuzzy Sets Syst. **159**(8), 990–1000 (2008)
4. Holland, J.H.: Adaptation in Natural and Artificial Systems. University of Michigan Press, Ann Arbor (1975)

# Chapter 2
# Theory and Background

In this chapter we present some basic concepts about the work to understand better the idea and the context of this book.

## 2.1 Fuzzy Logic

Fuzzy logic can be conceptualized as a generalization of classical logic. Modern fuzzy logic was developed by Lotfi Zadeh in the mid-1960s to model those problems in which imprecise data must be used or in which the rules of inference are formulated in a very general way making use of fuzzy categories. In fuzzy logic, which is also sometimes called multiple-valued logic, there are not just two alternatives but a whole continuum of truth values for logical propositions. A proposition A can have the truth value of 0.4 and its complement $A^c$ the truth value of 0.5. According to the type of negation operator that is used, the two truth values must not be necessarily add up to 1. Fuzzy logic has a weak connection to probability theory. Probabilistic methods that deal with imprecise knowledge are formulated in the Bayesian framework, but fuzzy logic does not need to be justified using a probabilistic approach. The common route is to generalize the findings of multi-valued logic in such a way as to preserve part of the algebraic structure [1–7].

Fuzzy set theory corresponds to fuzzy logic and the semantic of fuzzy operators can be understood using a geometric model. The geometric visualization of fuzzy logic will give us a hint as to the possible connection with neural networks. Fuzzy logic can be used as an interpretation model for the properties of neural networks, as well as for giving a more precise description of their performance. Fuzzy logic can also be used to specify networks directly without having to apply a learning algorithm. An expert in a certain field can sometimes produce a simple set of control rules for a dynamical system with less effort than the work involved in training a neural network. A classical example proposed by Zadeh to the neural network community is developing a system to park a car. It is straightforward to formulate a set of fuzzy rules for this task, but it is not immediately obvious how to build a network to do the same or how to train it. Fuzzy logic is now being used in

© The Author(s) 2016
L. Cervantes and O. Castillo, *Hierarchical Type-2 Fuzzy Aggregation of Fuzzy Controllers*, SpringerBriefs in Computational Intelligence,
DOI 10.1007/978-3-319-26671-8_2

many products of industrial and consumer electronics for which a good control system is sufficient and where the question of optimal control does not necessarily arise [8–13].

## 2.1.1   Fuzzy Set

The difference between crisp (i.e., classical) and fuzzy sets is established by introducing a membership function. Consider a finite set $X = \{x_1, x_2 \ldots x_n\}$ which will be considered the universal set in what follows. The subset A of $X$ consisting of the single element $x_1$ can be described by the n-dimensional membership vector Z (A) = (1, 0, 0 … 0), where the convention has been adopted that a 1 at the i-th position indicates that $x_i$ belongs to A. The set B composed of the elements $x_1$ and $x_n$ is described by the vector Z(B) = (1, 0, 0 … 1). Any other crisp subset of $X$ can be represented in the same way by an n-dimensional binary vector. But what happens if we lift the restriction to binary vectors? In that case we can define the fuzzy set C with the following vector description: $Z(C)$ = (0.5, 0, 0 … 0) [14–18].

In classical set theory such a set cannot be defined. An element belongs to a subset or it does not. In the theory of fuzzy sets we make a generalization and allow descriptions of this type. In our example the element $x_1$ belongs to the set C only to some extent. The degree of membership is expressed by a real number in the interval [0, 1], in this case 0.5. This interpretation of the degree of membership is similar to the meaning we assign to statements such as "person $x_1$ is an adult". Obviously, it is not possible to define a definite age which represents the absolute threshold to enter into adulthood. The act of becoming mature can be interpreted as a continuous process in which the membership of a person to the set of adults goes slowly from 0 to 1. There is many other examples of such diffuse statements. The concepts "old" and "young" or the adjectives "fast" and "slow" are imprecise but easy to interpret in a given context. In some applications, such as expert systems, for example, it is necessary to introduce formal methods capable of dealing with such expressions so that a computer using rigid Boolean logic can still process them. This is what the theory of fuzzy sets and fuzzy logic tries to accomplish [19–23].

## 2.1.2   Fuzzy Logic and Control

A fuzzy controller is a regulating system whose modus operandi is specified with fuzzy rules. In general it uses a small set of rules. The measurements are processed in their fuzzified form, fuzzy inferences are computed, and the result is defuzzified, that is, it is transformed back into a specific real number. The example of the electrical heater is explained to understand the idea in control and fuzzy logic: Assume that the room temperature is a number between 0 and 40 °C.

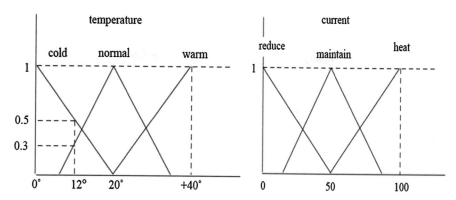

**Fig. 2.1** Membership functions for temperature and electric current categories

The controller can vary the electrical power consumed between 0 and 100 (in some suitable units), whereby 50 is the normal stand-by value. The temperature of $12°$ corresponds to the fuzzy number T = cold/0.5 + normal/0.3 + warm/0.0. These values lead to the previously computed inference action = heat/0.5 + maintain/0.3 + reduce/0.0. Membership degree of 0.5 represent a set of cold temperature, membership degree of 0.3 represent a set of normal temperature and membership degree of 0.0 represent a set of warm temperature. The controller must transform the result of this fuzzy inference into a definite value. The surfaces of the membership triangles below and the inferred degree of membership are calculated (see Fig. 2.1).

A fuzzy controller is just a system for the rapid computation of an approximation of a coarsely defined control surface. The fuzzy controller computes a control variable according to the values of the variables $x$ and $y$. Both variables are transformed into fuzzy categories. Assume that each variable is transformed into a combination of three categories. There are nine different combinations of the categories for x and y. For each of these nine combinations the value of the control variable is defined using a fuzzy rule [24, 25].

## 2.2  Granularity

Information granulation has emerged as one of the fundamental concepts of information processing giving rise to the discipline of Granular Computing. The concept itself permeates through a large variety of information systems the underlying idea is intuitive and appeals to our commonsense reasoning. We perceive the world by structuring our knowledge, perceptions, and acquired evidence in terms of information granules-entities, which are abstractions of the complex word and phenomena. By being abstract constructs, information granules and their ensuing processing done under the umbrella of Granular Computing, provides a conceptual and algorithmic framework to deal with an array of decision-making, control, and prediction problems. Granular Computing supports human-centric processing, which becomes an inherent feature of intelligent systems. The required level of detail becomes conveniently controlled by making suitable adjustments to the size of information granules and their distribution in the space in which the problem at hand is being positioned and handled. In this sense, an inherent flexibility, which comes hand in hand with Granular Computing, can be effectively exploited in various ways. Three general tendencies encountered in Granular Computing can be identified: (a) a design of information granules of higher order, (b) a development of information granules of higher type, and (c) a formation of hybrid information granules.

The basic concepts of graduation and granulation form the core of fuzzy logic, and are the main distinguishing features of fuzzy logic. More specifically, in fuzzy logic everything is or is allowed to be graduated, i.e., be a matter of degree or, equivalently, fuzzy. Furthermore, in fuzzy logic everything is or is allowed to be granulated, with a granule being a clump of attribute values drawn together by in distinguishability, similarity, proximity, or functionality. The concept of a generalized constraint serves to treat a granule as an object of computation. Graduated granulation, or equivalently fuzzy granulation, is a unique feature of fuzzy logic. Graduated granulation is inspired by the way in which humans deal with complexity and imprecision. The concepts of graduation, granulation, and graduated granulation play key roles in granular computing. Graduated granulation underlies the concept of a linguistic variable, i.e., a variable whose values are words rather than numbers. In retrospect, this concept, in combination with the associated concept of a fuzzy if–then rule, may be viewed as a first step toward granular computing.

Granular Computing (GrC) is a general computation theory for effectively using granules such as subsets, neighborhoods, ordered subsets, relations (subsets of products), fuzzy sets (membership functions), variables (measurable functions), Turing machines (algorithms), and intervals to build an efficient computational model for complex with huge amounts of data, information and knowledge [26, 27].

## 2.3  Genetic Algorithms

Genetic algorithms (GAs) were proposed by John Holland in the 1960s and were developed by Holland and his students and colleagues at the University of Michigan in the 1960s and the 1970s [8]. In contrast with evolution strategies and evolutionary programming, Holland's original goal was not to design algorithms to solve specific problems, but rather to formally study the phenomenon of adaptation as it occurs in nature and to develop ways in which the mechanisms of natural adaptation might be imported into computer systems. Holland's 1975 book Adaptation in Natural and Artificial Systems presented the genetic algorithm as an abstraction of biological evolution and gave a theoretical framework for adaptation under the GA. Holland's GA is a method for moving from one population of "chromosomes" (e.g., strings of ones and zeros, or "bits") to a new population by using a kind of "natural selection" together with the genetics—inspired operators of crossover, mutation, and inversion. Each chromosome consists of "genes" (e.g., bits), each gene being an instance of a particular "allele" (e.g., 0 or 1). The selection operator chooses those chromosomes in the population that will be allowed to reproduce, and on average the fitter chromosomes produce more offspring than the less fit ones. Crossover exchanges subparts of two chromosomes, roughly mimicking biological recombination between two single—chromosome ("haploid") organisms; mutation randomly changes the allele values of some locations in the chromosome; and inversion reverses the order of a contiguous section of the chromosome, thus rearranging the order in which genes are arrayed. (Here, as in most of the GA literature, "crossover" and "recombination" will mean the same thing.)

Holland's introduction of a population—based algorithm with crossover, inversion, and mutation was a major innovation. (Rechenberg's evolution strategies started with a "population" of two individuals, one parent and one offspring, the offspring being a mutated version of the parent; many—individual populations and crossover were not incorporated until later. Fogel, Owens, and Walsh's evolutionary programming likewise used only mutation to provide variation.) Moreover, Holland was the first to attempt to put computational evolution on a firm theoretical footing [28]. Until recently this theoretical foundation, based on the notion of "schemas," was the basis of almost all subsequent theoretical work on genetic algorithms [29–32]. Genetic algorithms (GAs) for rule discovery can be divided into two broad approaches, based on how rules are encoded in the population of individuals ("chromosomes"). In the Michigan approach each individual encodes a single prediction rule, whereas in the Pittsburgh approach each individual encodes a set of prediction rules [8].

## 2.4   Aviation

There are four fundamental basic flight maneuvers upon which all flying tasks are based: straight and level flight, turns, climbs, and descents. All controlled flight consists of either one, or a combination or more than one, of these basic maneuvers. If a student pilot is able to perform these maneuvers well, and the student's proficiency is based on accurate "feel" and control analysis rather than mechanical movements, the ability to perform any assigned maneuver will only be a matter of obtaining a clear visual and mental conception of it. The pilot should always be considered the center of movement of the airplane, or the reference point from which the movements of the airplane are judged and described [33–37]. The following will always be true, regardless of the airplane's attitude in relation to the Earth:

- When back pressure is applied to the elevator control, the airplane's nose rises in relation to the pilot.
- When forward pressure is applied to the elevator control, the airplane's nose lowers in relation to the pilot.
- When right pressure is applied to the aileron control, the airplane's right wing lowers in relation to the pilot.
- When left pressure is applied to the aileron control, the airplane's left wing lowers in relation to the pilot.
- When pressure is applied to the right rudder pedal, the airplane's nose moves (yaws) to the right in relation to the pilot.
- When pressure is applied to the left rudder pedal, the airplane's nose moves (yaws) to the left in relation to the pilot.

The preceding explanations should prevent the beginning pilot from thinking in terms of "up" or "down" in respect to the Earth, which is only a relative state to the pilot. It will also make understanding of the functions of the controls much easier, particularly when performing steep banked turns and the more advanced maneuvers. Consequently, the pilot must be able to properly determine the control application required to place the airplane in any attitude or flight condition that is desired. The flight instructor should explain that the controls will have a natural "live pressure" while in flight and that they will remain in neutral position of their own accord, if the airplane is trimmed properly. With this in mind, the pilot should be cautioned never to think of movement of the controls, but of exerting a force on them against this live pressure or resistance. Movement of the controls should not be emphasized; it is the duration and amount of the force exerted on them that effects the displacement of the control surfaces and maneuvers the airplane. The amount of force the airflow exerts on a control surface is governed by the airspeed and the degree that the surface is moved out of its neutral or streamlined position.

Since the airspeed will not be the same in all maneuvers, the actual amount the control surfaces are moved is of little importance; but it is important that the pilot maneuver the airplane by applying sufficient control pressure to obtain a desired result, regardless of how far the control surfaces are actually moved. The controls should be held lightly, with the fingers, not grabbed and squeezed. Pressure should be exerted on the control yoke with the fingers. A common error in beginning pilots is a tendency to "choke the stick." This tendency should be avoided as it prevents the development of "feel," which is an important part of aircraft control. The pilot's feet should rest comfortably against the rudder pedals. Both heels should support the weight of the feet on the cockpit floor with the ball of each foot touching the individual rudder pedals. The legs and feet should not be tense; they must be relaxed just as when driving an automobile [38–40].

## 2.4.1   Attitude Flying

Attitude is the angular difference measured between an airplane's axis and the line of the Earth's horizon. Pitch attitude is the angle formed by the longitudinal axis, and bank attitude is the angle formed by the lateral axis. Rotation about the airplane's vertical axis (yaw) is termed an attitude relative to the airplane's flight path, but not relative to the natural horizon. In attitude flying, airplane control is composed of four components: pitch control, bank control, power control, and trim [41].

- Pitch control is the control of the airplane about the lateral axis by using the elevator to raise and lower the nose in relation to the natural horizon.
- Bank control is control of the airplane about the longitudinal axis by use of the ailerons to attain a desired bank angle in relation to the natural horizon.
- Power control is used when the flight situation indicates a need for a change in thrust.
- Trim is used to relieve all possible control pressures held after a desired attitude has been attained.

## 2.4.2   Level Turns

A turn is made by banking the wings in the direction of the desired turn. A specific angle of bank is selected by the pilot, control pressures applied to achieve the desired bank angle, and appropriate control pressures exerted to maintain the desired bank angle once it is established. All four primary controls are used in close coordination when making turns. Their functions are as follows:

- The ailerons bank the wings and so determine the rate of turn at any given airspeed.
- The elevator moves the nose of the airplane up or down in relation to the pilot, and perpendicular to the wings. Doing that, it both sets the pitch attitude in the turn and "pulls" the nose of the airplane around the turn.
- The throttle provides thrust, which may be used for airspeed to tighten the turn.
- The rudder offsets any yaw effects developed by the other controls. The rudder does not turn the airplane.

  For purposes of this discussion, turns are divided into three classes: shallow turns, medium turns, and steep turns:

- Shallow turns are those in which the bank (less than approximately 20°) is so shallow that the inherent lateral stability of the airplane is acting to level the wings unless some aileron is applied to maintain the bank.
- Medium turns are those resulting from a degree of bank (approximately 20°–45°) at which the airplane remains at a constant bank.

- Steep turns are those resulting from a degree of bank (45° or more) at which the "overbanking tendency" of an airplane overcomes stability, and the bank increases unless aileron is applied to prevent it [42, 43].

# References

1. Cervantes, L., Castillo, O.: Design of a fuzzy system for the longitudinal control of an F-14 airplane. In: Soft Computing for Intelligent Control and Mobile Robotics, vol. 318, pp. 213–224. Springer, Berlin (2011)
2. Cervantes, L., Castillo, O.: Intelligent control of nonlinear dynamic plants using a hierarchical modular approach and type-2 fuzzy logic. In: Lecture Notes in Computer Science, vol. 7095, pp. 1–12. Springer, Berlin (2011)
3. Cervantes, L., Castillo, O.: Hierarchical genetic algorithms for optimal type-2 fuzzy system design. In: Annual Meeting of the North American Fuzzy Information Processing Society, pp. 324–329 (2011)
4. Cervantes, L., Castillo, O.: Automatic design of fuzzy systems for control of aircraft dynamic systems with genetic optimization. In: World Congress and AFSS International Conference, pp. OS-413-1–OS-413-7 (2011)
5. Cervantes, L., Castillo, O.: Comparative study of type-1 and type-2 fuzzy systems for the three-tank water control problem. In: LNAI, vol. 7630, pp. 362–373. Springer, Berlin (2013)
6. Cervantes, L., Castillo, O.: Genetic design of optimal type-1 and type-2 fuzzy systems for longitudinal control of an airplane. J. Intell. Autom. Soft Comput. 20(2), 213–227 (2014)
7. Chalupa, P., Novák, J., Bobál, V.: Detailed Simulink model of real time three tank system. In: Proceedings of the 2nd International Conference on Circuits, Systems, Communications and Computers 2011, CSCC'11, pp. 161–166 (2011)
8. Holland, J.H.: Adaptation in Natural and Artificial Systems. University of Michigan Press, Ann Arbor (1975)
9. Lam, H., Leung, F.: Stability analysis of discrete-time fuzzy-model-based control systems with time delay: time delay-independent approach. Fuzzy Sets Syst. 159(8), 990–1000 (2008)
10. Lam, H., Leung, F.: Fuzzy rule-based combination of linear and switching state-feedback controllers. Fuzzy Sets Syst. 156(2), 153–184 (2005)
11. Li, I.-H., Lee, L.-W.: A hierarchical structure of observer-based adaptive fuzzy-neural controller for MIMO systems. Fuzzy Sets Syst. 185(1), 52–82 (2011)
12. Malla, S., Bhende, C.: Voltage control of stand-alone wind and solar energy system. Int. J. Electr. Power Energy Syst. 56, 361–373 (2014)
13. Niemann, H., Stoustrup, J.: Passive fault tolerant control of a double inverted pendulum a case study. Control Eng. Pract. 13(8), 1047–1059 (2005)
14. Oh, S.-K., Jung, S.-H., Pedrycz, W.: Design of optimized fuzzy cascade controllers by means of hierarchical fair competition-based genetic algorithms. Expert Syst. Appl. 36(9), 11641–11651 (2009)
15. Ornelas-Tellez, F., Sanchez, E., Loukianov, A., Rico, J.: Robust inverse optimal control for discrete-time nonlinear system stabilization. Eur. J. Control 20(1), 38–44 (2014)
16. Sung, H., Kim, D., Park, J., Joo, Y.: Robust digital control of fuzzy systems with parametric uncertainties: LMI-based digital redesign approach. Fuzzy Sets Syst. 161(6), 919–933 (2010)
17. Warren, P.: Mechanics of Flight, 2nd edn. Wiley, Hoboken, New Jersey (2010)
18. Castillo, O., Melin, P.: A review on the design and optimization of interval type-2 fuzzy controllers. Appl. Soft Comput. 12(4), 1267–1278 (2012)
19. Castillo, O., Melin, P.: New fuzzy-fractal-genetic method for automated mathematical modelling and simulation of robotic dynamic systems. In: IEEE International Conference on Fuzzy Systems, vol. 2, pp. 1182–1187 (1998)

20. Chen, G., Pham, T.: Introduction to Fuzzy Sets, Fuzzy Logic, and Fuzzy Control Systems. CRC Press, Boca Raton (2001)
21. Dadios, E.: Fuzzy Logic-Controls, Concepts, Theories and Applications (2012)
22. Sefer, K., Omer, C., Okyay, K.: Adaptive neuro-fuzzy inference system based autonomous flight control of unmanned air vehicles. Expert Syst. Appl. J. **37**(2), 1229–1234 (2010)
23. Sepulveda, R., Castillo, O., Melin, P., Montiel, O.: An efficient computational method to implement type-2 fuzzy logic in control application. In: Analysis and Design of Intelligent System Using Soft Computing Techniques 2007, pp. 45–52 (2007)
24. Zadeh, L.: Fuzzy Sets and Fuzzy Information Granulation Theory. Beijing Normal University Press, Beijing (2000)
25. Zadeh, L.A.: Some reflections on soft computing, granular computing and their roles in the conception, design and utilization of information/intelligent systems. Soft Comput. **2**, 23–25 (1998)
26. Yu, Y., Chen, L., Sun, F., Wu, C.: Matlab/Simulink-based simulation for digital-control system of marine three-shaft gas-turbine. Appl. Energy **80**(1), 1–10 (2005)
27. Ouyang, P.R., Acob, J., Pano, V.: PD with sliding mode control for trajectory tracking of robotic system. Robot. Comput. Integr. Manuf. **30**(2), 189–200 (2014)
28. Castillo, O., Martinez-Marroquin, R., Melin, P., Valdez, F., Soria, J.: Comparative study of bio-inspired algorithms applied to the optimization of type-1 and type-2 fuzzy controllers for an autonomous mobile robot. Inf. Sci. **192**, 19–38 (2012)
29. Cázarez, N., Aguilar, L., Castillo, O.: Fuzzy logic control with genetic parameters optimization for the output regulation of a servomechanism with nonlinear backlash. Expert Syst. Appl. **37**(6), 4368–4378 (2010)
30. Haupt, R., Haupt, S.: Practical Genetic Algorithm. Wiley Interscience, Hoboken (2004)
31. Hidalgo, D., Melin, P., Castillo, O.: An optimization method for designing type-2 fuzzy inference systems based on the footprint of uncertainty using genetic algorithms. Expert Syst. Appl. **39**(4), 4590–4598 (2012)
32. No, T.S., Mina, B.M., Stone, R.H., Wong, K.C.: Control and Simulation of Arbitrary Flight Trajectory-Tracking, pp. 560–756. Department of Aerospace Engineering, Chonbuk National University, Deokjin Dong, Chonju NSW (2006)
33. Caughey, D.: Introduction of Aircrafts Stability and Control Course Notes for M&AE 5070. Sibbley School of Mechanical and Aerospace Engineering, Cornell University, New York, pp. 14853–17501 (2011)
34. No, T.S., Kim, J.-E., Moon, J.H., Kim, S.J.: Modeling, control, and simulation of dual rotor wind turbine generator system. Renew. Energy **34**(10), 2124–21322 (2009)
35. Pedrycz, W., Chen, S.: Granular Computing and Intelligent System, Design with Information Granules of Higher Order and Higher Type. Intelligent Systems Reference Library, vol. 13 (2011)
36. Rachman, E., Jaam, J., Hasnah, A.: Non-linear simulation of controller for longitudinal control augmentation system of F-16 using numerical approach. Inf. Sci. J. **164**(1–4), 47–60 (2004)
37. Sanchez, E., Becerra, H., Velez, C.: Combining fuzzy, PID and regulation control for an autonomous mini-helicopter. J. Inf. Sci. **177**(10), 1999–2022 (2007)
38. Schmidt, D.: Modern Flight Dynamics. McGraw Hill, New York (2012)
39. Song, Q., Song, Y.D.: Generalized PI control design for a class of unknown nonaffine systems with sensor and actuator faults. Syst. Control Lett. **64**, 86–95 (2014)
40. Song, Y., Wang, H.: Design of flight control system for a small unmanned tilt rotor aircraft. Chin. J. Aeronaut. **22**(3), 250–256 (2009)
41. Tao, C., Taur, J., Wang, C., Chen, U.: Fuzzy hierarchical swing-up and sliding position controller for the inverted pendulum–cart system. Fuzzy Sets Syst. **159**(20), 2763–2784 (2008)
42. Zadeh, L.A.: Outline of a new approach to the analysis of complex systems and decision processes. IEEE Trans. Syst. ManCybern. SMC **3**, 28–44 (1973)
43. Zhang, Y., Wang, Q., Dong, Ch., Jiang, Y.: H∞ output tracking control for flight control systems with time-varying delay. Chin. J. Aeronaut. **26**(5), 1251–1258 (2013)

# Chapter 3
# Control Problem and Proposed Method

In this chapter the proposed method is illustrated and the control problems are explained.

## 3.1 Proposed Method

Existing control problems can be solved using fuzzy systems [1], or some other technique [2–6] that enables us to have a good control on the problem [7–12]. Usually Mamdani fuzzy models are used.

In this book, a new hierarchical approach combining multiple controllers is used as a new method for global control. In particular, for a complex control problem that can be solved by dividing it into several simple controllers, then and the architecture shown in Figs. 3.1 and 3.2 can be used. In this case each individual fuzzy controller solves only a part of the control problem, but the overall result is the control of the complex plant (see Figs. 3.1 and 3.2).

Figure 3.1 represents a control problem with a single output, this case is very common when the problem is simple and has one output, and in that case the input is going to the fuzzy system and the output goes to the simulation plant. In Fig. 3.2 the architecture of a problem with multiple controllers is presented, where it is possible to have one or more inputs and these inputs going to the individual fuzzy systems, then the outputs go to the simulation plant and then the feedback is carried out.

When the problem of control is selected and the fuzzy system is used to achieve the goal of control, rules are necessary in the fuzzy system and those rules can be designed with different methods. One of these methods is by using adaptive neuro-fuzzy inference systems (ANFIS) to design the rules, another method is by using an expert system, but when the user of the plant has the theoretical knowledge about the operation of the problem then the rules can be designed empirically considering the knowledge and literature of the case of study. The error can be calculated with different methods such as the sum of squared error, absolute error and relative error, in this book we use the absolute error, we decide to calculate individual errors in both cases because in the first case of study we established a desire level in each water tanks, in the second case of study when a trajectory of

© The Author(s) 2016
L. Cervantes and O. Castillo, *Hierarchical Type-2 Fuzzy Aggregation
of Fuzzy Controllers*, SpringerBriefs in Computational Intelligence,
DOI 10.1007/978-3-319-26671-8_3

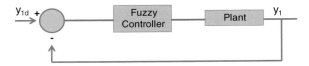

**Fig. 3.1** Architecture of a single controller

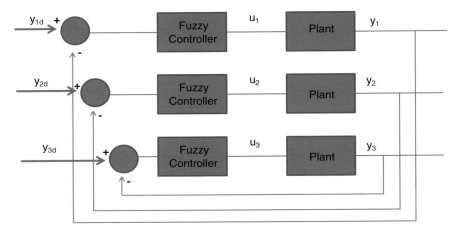

**Fig. 3.2** Architecture using multiple controllers

flight is established the elevator, aileron and the rudder have values individually depends on the trajectory (height, wind, etc.) and we evaluate individually each controller. Also the individual errors can be added for an average final error, but in this case we want to obtain individual errors and observe the behavior of each controller. In this work the architecture of the conventional controller is changed with a new module for aggregation in which the individual controller results are combined with a fuzzy system as an aggregator of the outputs. The proposed method when this new aggregation module is applied is shown in Fig. 3.3

Today there are more complex problems that are difficult to control and it is for this reason that in this book a modular approach based on granularity [13, 14] and fuzzy logic is proposed and shown in Fig. 3.4. This figure represents the proposed method but Figs. 3.4 and 3.5 is otherwise to represent the proposed method but is the same idea. In this case, a genetic algorithm is used to optimize the design of the modular controller.

The main contribution of this work is the fuzzy granular models to improve the solution of the control problem that is going to be considered, since it divides the problem in modules for the different types of control and this model will receive the signal for further processing and perform adequate control. This architecture is useful in many cases to develop each controller separately. In Fig. 3.4 an example is presented and shows how this architecture can be used in the control area. In this

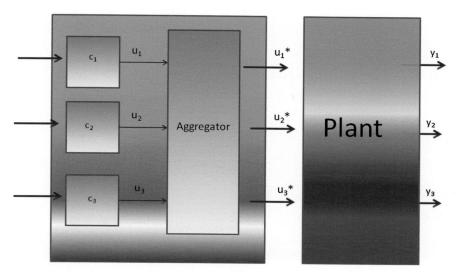

**Fig. 3.3**  Proposed controller using an aggregator

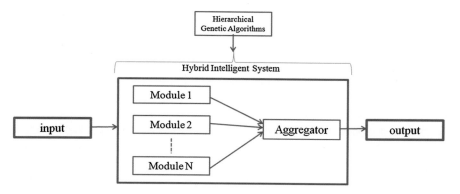

**Fig. 3.4**  Proposed modular approach for control

figure input 1 represents all the data of the all controllers that can have the case of study, then the controllers are divided to apply the aggregator, in Fig. 3.5 we can see that the first part is represent for Fig. 3.4 where the input and the output is the set of data in second part of the Fig. 3.5 we illustrate that the input can be have more than one input in this case complex problems with more than 2 controllers and also we can have more than 1 outputs. Is important to mention that each module has different data, each module contained the data of one controller. In this example the fuzzy logic controller has inputs 1 to n and outputs are also 1 to n. When more than one controller is needed for the case of study, type-1 fuzzy logic in each controller can be used and then the outputs of the type-1 fuzzy systems will be taken as inputs

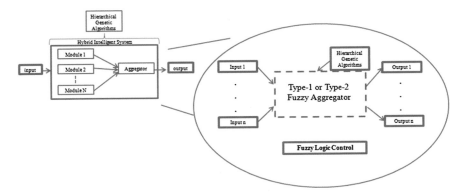

**Fig. 3.5** Proposed granular fuzzy system

in the type-1 or type-2 fuzzy aggregator system to obtain new outputs (see Fig. 3.5). The granularity in this book is used when we divides the problem in sub-problems and are add to the aggregator and then granulate the outputs.

## 3.2  Problem Description

There are several methods available to solve control problems such as: PID control, fuzzy control and some other methods using neural networks [15–24]. In this work the main goal is to apply the proposed method in automatic control of nonlinear dynamic plants using a fuzzy granular approach and also to implement a granular architecture in a particular control problem. In this work we use two cases of control to prove the architecture and these control problems are explained in Sects. 3.2.1 and 3.2.2.

### 3.2.1  First Complex Control Problem

To implement the proposed method the case of study needs to have more than one individual controller, and the control of the 3 water tanks system is the first case of study. The 3 tanks include valves that are open or closed; these valves must be well controlled to produce the desired level of water in each of the three tanks. Tank 1 and tank 3 have a valve that provides water(valve 1 and valve 2), between tank 1 and tank 2 is one valve (valve 13), valve 32 is between tank 2 and tank 3, finally tank 3 has one valve to lets water drain out. All valves need to control the opening and closing to provide water and obtain the desired level in all the tanks. In this

**Fig. 3.6** Water control of 3 tanks

control problem 5 controllers are use (valve 1, valve 13, valve 32, valve 20 and valve 2), in this case 5 fuzzy systems are necessary to control all valves. The control problem is illustrated in Fig. 3.6.

### 3.2.2   Second Complex Control Problem

The second benchmark used to prove the proposed method was the flight control, this case of study is to control the flight of an airplane, first a brief explanation about the problem: In this problem we need to control the total flight and for this reason is necessary to know that to control the flight is necessary 3 controllers (longitudinal, lateral and direction control) [15, 21]. For longitudinal control is important to consider the stick of the pilot because with this the elevator are moved (up and down pulling or pushing the stick) and this movement change the position of the nose of the airplane (movement in axes $y$). To direction control, the pedals make changes when pedals are press (left or right) the rudder move (left or right) and the aircraft is moving left or right in axes $x$. Finally the lateral control change moving the stick left or right when this happened the aileron are move up or down (alternating) and the aircraft is move in axes $z$. Based on explanation above we decide to perform first the individual controllers (Longitudinal, lateral and direction control).

A fuzzy system was designed for each controller. To obtain the longitudinal control a fuzzy system with 1 input and 1 output was developed, in the output variable is elevator because when it moves the aircraft moves the y position and in the input the Stick is used, in this the linguistic variables are "push", "pull" or "not movel" the stick and depending on the movement a reaction in the elevator is obtain.

In the lateral control the outputs are the ailerons because when the ailerons are moved the aircraft can change the direction in axes $z$ and in the input are the stick of the pilot and when this is move (right or left) the ailerons have a reaction (up/down)

**Fig. 3.7** Movements of the airplane

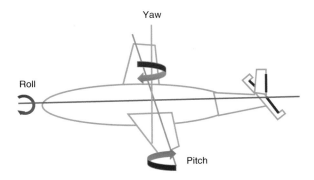

and the linguistic variables are stick (move right, move left, center) (see Fig. 3.7). In direction control the output is the rudder when the rudder is move (left or right) the aircraft change the trajectory in axes *x*, in the input the pedals are consider (left and right). When pedals are pressed the rudder has a reaction to left or right and the aircraft change the position to the new direction.

# References

1. Li, I.-H., Lee, L.-W.: A hierarchical structure of observer-based adaptive fuzzy-neural controller for MIMO systems. Fuzzy Sets Syst. **185**(1), 52–82 (2011)
2. Ornelas-Tellez, F., Sanchez, E., Loukianov, A., Rico, J.: Robust inverse optimal control for discrete-time nonlinear system stabilization. Eur. J. Control **20**(1), 38–44 (2014)
3. Ouyang, P.R., Acob, J., Pano, V.: PD with sliding mode control for trajectory tracking of robotic system. Rob. Comput. Integr. Manuf. **30**(2), 189–200 (2014)
4. Pedrycz, W., Chen, S.: Granular computing and intelligent system, design with information granules of higher order and higher type. In: Intelligent Systems Reference Library, vol. 13 (2011)
5. Rachman, E., Jaam, J., Hasnah, A.: Non-linear simulation of controller for longitudinal control augmentation system of F-16 using numerical approach. Inf. Sci. J. **164**(1–4), 47–60 (2004)
6. Sanchez, E., Becerra, H., Velez, C.: Combining fuzzy, PID and regulation control for an autonomous mini-helicopter. J. Inf. Sci. **177**(10), 1999–2022 (2007)
7. Sefer, K., Omer, C., Okyay, K.: Adaptive neuro-fuzzy inference system based autonomous flight control of unmanned air vehicles. Expert Syst. Appl. J. **37**(2), 1229–1234 (2010)
8. Song, Y., Wang, H.: Design of flight control system for a small unmanned tilt rotor aircraft. Chin. J. Aeronaut. **22**(3), 250–256 (2009)
9. Tao, C., Taur, J., Wang, C., Chen, U.: Fuzzy hierarchical swing-up and sliding position controller for the inverted pendulum–cart system. Fuzzy Sets Syst. **159**(20), 2763–2784 (2008)
10. Warren, P.: Mechanics of Flight, 2nd edn. Wiley, Hoboken, New Jersey (2010)
11. Yu, Y., Chen, L., Sun, F., Wu, C.: Matlab/Simulink-based simulation for digital-control system of marine three-shaft gas-turbine. Appl. Energy **80**(1), 1–10 (2005)
12. Zadeh, L.A.: Some reflections on soft computing, granular computing and their roles in the conception, design and utilization of information/intelligent systems. Soft Comput. **2**, 23–25 (1998)
13. Sung, H., Kim, D., Park, J., Joo, Y.: Robust digital control of fuzzy systems with parametric uncertainties: LMI-based digital redesign approach. Fuzzy Sets Syst. **161**(6), 919–933 (2010)

14. Zadeh, L.: Fuzzy Sets and Fuzzy Information Granulation Theory. Beijing Normal University Press, Beijing (2000)
15. Dadios, E.: Fuzzy Logic-Controls, Concepts, Theories and Applications (2012)
16. Niemann, H., Stoustrup, J.: Passive fault tolerant control of a double inverted pendulum a case study. Control Eng. Pract. **13**(8), 1047–1059 (2005)
17. Lam, H., Leung, F.: Stability analysis of discrete-time fuzzy-model-based control systems with time delay: time delay-independent approach. Fuzzy Sets Syst. **159**(8), 990–1000 (2008)
18. No, T.S., Mina, B.M., Stone, R.H., Wong, K.C.: Control and Simulation of Arbitrary Flight Trajectory-Tracking, pp. 560–756. Department of Aerospace Engineering, Chonbuk National University, Deokjin Dong, Chonju NSW (2006)
19. Lam, H., Leung, F.: Fuzzy rule-based combination of linear and switching state-feedback controllers. Fuzzy Sets Syst. **156**(2), 153–184 (2005)
20. Holland, J.H.: Adaptation in Natural and Artificial Systems. University of Michigan Press, Ann Arbor (1975)
21. Haupt, R., Haupt, S.: Practical Genetic Algorithm. Wiley Interscience, Hoboken (2004)
22. Cervantes, L., Castillo, O.: Intelligent control of nonlinear dynamic plants using a hierarchical modular approach and type-2 fuzzy logic. In: Lecture Notes in Computer Science, vol. 7095, pp. 1–12. Springer, Berlin (2011)
23. Cervantes, L., Castillo, O.: Hierarchical genetic algorithms for optimal type-2 fuzzy system design. In: Annual Meeting of the North American Fuzzy Information Processing Society, pp. 324–329 (2011)
24. Cervantes, L., Castillo, O.: Automatic design of fuzzy systems for control of aircraft dynamic systems with genetic optimization. In: World Congress and AFSS International Conference, pp. OS-413-1–OS-413-7 (2011)

# Chapter 4
# Simulation Results

In this chapter simulation results are presented using the proposed method and the two cases of control. In each case of study the fuzzy systems are used and genetic algorithms are implemented to improve the results.

## 4.1 Simulation Results in the First Case of Study

In this case study is necessary to control the openness and closeness for the 5 valves and each valve needs a fuzzy system to control the openness and closeness. In this case 5 fuzzy systems are necessary to control all the valves, one fuzzy system for each valve. The inputs of the fuzzy system are (tank 1, tank 2 and tank 3), and the outputs are (valve 1, valve 13, valve 32, valve 20 and valve 2). The first fuzzy system has one input (water level of tank 1) and the output is valve 1, the second fuzzy system has 2 inputs (tank 1 and tank 2) and the output is valve 13, next fuzzy system has 2 inputs (tank 2 and tank 3) and the output is valve 32, the fourth fuzzy system has one input (tank 3) and the output is valve 2, and finally the last fuzzy system has one input (tank 3) and output valve 20. Is important to know that when valves are between 2 tanks is necessary use both inputs (tanks) to control the opening and closing of each valve. The fuzzy systems used are shown in Figs. 4.1, 4.2, 4.3, 4.4 and 4.5.

Having obtained the fuzzy systems the simulation was performed using the Matlab programming language and simulation. In this plant the 5 fuzzy systems were included and then the simulation was performed. In the fuzzy systems, triangular, Gaussian and trapezoidal membership function were used to perform the simulations. The simulation plant used in this case of study is illustrated in Fig. 4.6.

Results of the simulation are shown in Table 4.1 using different types of membership function in each valve.

Previous results were obtained using a type-1 fuzzy system and their rules are shown in Figs. 4.7, 4.8, 4.9, 4.10 and 4.11.

The fuzzy rules presented above were used to control the valves, and in some controllers is necessary to have up to have 9 rules. Rules depend on the inputs and the outputs of the controller and also the number of the membership functions. In this case, in all the controllers the inputs and the outputs have 3 membership functions.

© The Author(s) 2016
L. Cervantes and O. Castillo, *Hierarchical Type-2 Fuzzy Aggregation of Fuzzy Controllers*, SpringerBriefs in Computational Intelligence,
DOI 10.1007/978-3-319-26671-8_4

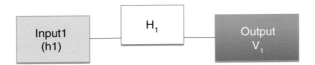

**Fig. 4.1** Fuzzy system to control valve 1

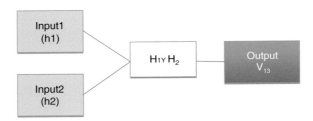

**Fig. 4.2** Fuzzy system to control valve 13

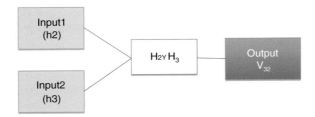

**Fig. 4.3** Fuzzy system to control valve 32

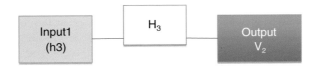

**Fig. 4.4** Fuzzy system to control valve 2

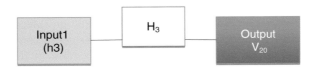

**Fig. 4.5** Fuzzy system to control valve 20

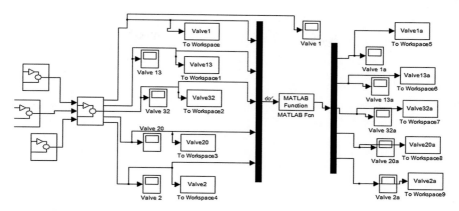

**Fig. 4.6**  Simulation plant

**Table 4.1**  Results for the simulation plant with type-1 fuzzy system using different types of membership function

| Valve | Error using Gaussian membership function | Error using trapezoidal membership function | Error using triangular membership function |
|-------|-------------------------------------------|-----------------------------------------------|----------------------------------------------|
| Valve 1 | 0.8980 | 0.9522 | 0.9246 |
| Valve 13 | 0.8994 | 0.9551 | 0.9278 |
| Valve 2 | 0.8994 | 0.9504 | 0.9278 |
| Valve 20 | 0.8995 | 0.9551 | 0.9279 |
| Valve 32 | 0.8463 | 0.8233 | 0.8341 |

1.  If (h1 is LOW) then (v1 is 1)
2.  If (h1 is NORMAL) then (v1 is 0)
3.  If (h1 is HIGH) then (v1 is m1)

**Fig. 4.7**  Rules of the fuzzy controller 1

**Fig. 4.8**  Rules of the fuzzy controller 2

1.  If (h1 is LOW) and (h3 is LOW) then (v13 is 1)
2.  If (h1 is LOW) and (h3 is NORMAL) then (v13 is 1)
3.  If (h1 is LOW) and (h3 is HIGH) then (v13 is 1)
4.  If (h1 is NORMAL) and (h3 is LOW) then (v13 is 0)
5.  If (h1 is NORMAL) and (h3 is NORMAL) then (v13 is 0)
6.  If (h1 is NORMAL) and (h3 is HIGH) then (v13 is 0)
7.  If (h1 is HIGH) and (h3 is LOW) then (v13 is m1)
8.  If (h1 is HIGH) and (h3 is NORMAL) then (v13 is m1)
9.  If (h1 is HIGH) and (h3 is HIGH) then (v13 is m1)

| 1. | If (h2 is LOW) then (v2 is 1) |
| 2. | If (h2 is NORMAL) then (v2 is 0) |
| 3. | If (h2 is HIGH) then (v2 is m1) |

**Fig. 4.9** Rules of the fuzzy controller 3

| 1. | If (h2 is LOW) then (v20 is m1) |
| 2. | If (h2 is NORMAL) then (v20 is 0) |
| 3. | If (h2 is HIGH) then (v20 is 1) |

**Fig. 4.10** Rules of the fuzzy controller 4

**Fig. 4.11** Rules of the fuzzy controller 5

| 1. | If (h2 is LOW) and (h3 is LOW) then (v32 is 1) |
| 2. | If (h2 is LOW) and (h3 is NORMAL) then (v32 is 1) |
| 3. | If (h2 is LOW) and (h3 is HIGH) then (v32 is 1) |
| 4. | If (h2 is NORMAL) and (h3 is LOW) then (v32 is 1) |
| 5. | If (h2 is NORMAL) and (h3 is NORMAL) then (v32 is 0) |
| 6. | If (h2 is NORMAL) and (h3 is HIGH) then (v32 is 1) |
| 7. | If (h2 is HIGH) and (h3 is LOW) then (v32 is 1) |
| 8. | If (h2 is HIGH) and (h3 is NORMAL) then (v32 is 1) |
| 9. | If (h2 is HIGH) and (h3 is HIGH) then (v32 is 1) |

Two of the five controllers have 2 inputs because they are between 2 tanks, for example valve 13 is between tank 1 and tank 2, and valve 32 is located between tank 2 and tank 3. In these cases nine fuzzy rules are necessary to control the valves, because the number of the inputs and the membership functions. In the other cases the controllers have one input and there are only 3 rules.

The proposed method is applied by using a type-1 fuzzy system to integrate the outputs of the individual valve fuzzy controllers. The granular type-1 fuzzy system has 5 inputs and 5 outputs. Previously, type-1 fuzzy systems were used and results were obtained, these outputs are used as new inputs to the granular type-1 fuzzy system, this is to obtain new outputs and to have improved results with this method. The granular type-1 fuzzy system is shown in Fig. 4.12.

In the aggregator above to obtain the rules we used a genetic algorithm using a Michigan approach because it is the classical approach, having proven itself and undergone more development and we wanted that each rule would represent by a gen, and 10 genes are considered to optimize the fuzzy rules, all the genes have 3 values (1, 2 and 3), for example gen 1 has values between 1 and 3 (1 is open, 2 medium and 3 is closed), for the first input. Each gene 1–10 is a variable representing 1–5 that are the input variables and 6–10 that are output variables. Each gene has values between 1 and 3 representing whether the valve is open or closed. Each generation of the genetic algorithm is changing the rules (antecedents and

**Fig. 4.12** Aggregator for the water control

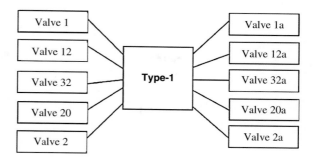

consequences). And in the end gets the best rules that are generated during all evolutions. The fitness function used is the one in Eq. 4.1.

$$f(Y) = \frac{\left( \sum_{i=1}^{n} \frac{|y_{REF1}^i|}{n} - \frac{|y_{FS1}^i|}{n} + \sum_{i=1}^{n} \frac{|y_{REF2}^i|}{n} - \frac{|y_{FS2}^i|}{n} + \sum_{i=1}^{n} \frac{|y_{REFN}^i|}{n} - \frac{|y_{FSN}^i|}{n} \right)}{N} \tag{4.1}$$

where $y_{REF}$ is the reference, $y_{FS}$ is the output of the controller and **n** is the number of points used in the comparison and N is the number of the controllers. The parameters used in the GA are: generations = 20, individuals = 25, selection = roulette, discrete Mutation, crossover of a single point with a crossover rate = 0.9 and mutation rate = 0.2.

After the genetic algorithm evolution is done the optimal set of fuzzy rules is the following:

1. If ($V_1$ is medium) and ($V_{13}$ is closed) and ($V_{32}$ is medium) and ($V_{20}$ is medium) and ($V_2$ is medium) then ($NV_1$ is medium) ($NV_{13}$ is medium) ($NV_{32}$ is closed) ($NV_{20}$ is medium) ($NV_2$ is 1)
2. If ($V_1$ is open) and ($V_{13}$ is medium) and ($V_{32}$ is medium) and ($V_{20}$ is closed) and ($V_2$ is open) then ($NV_1$ is medium) ($NV_{13}$ is medium) ($NV_{32}$ is open) ($NV_{20}$ is medium) ($NV_2$ is medium)
3. If ($V_1$ is closed) and ($V_{13}$ is closed) and ($V_{32}$ is medium) and ($V_{20}$ is open) and ($V_2$ is medium) then ($NV_1$ is open) ($NV_{13}$ is open) ($NV_{32}$ is medium) ($NV_{20}$ is open) ($NV_2$ is open)
4. If ($V_1$ is open) and ($V_{13}$ is closed) and ($V_{32}$ is closed) and ($V_{20}$ is medium) and ($V_2$ is closed) then ($NV_1$ is closed) ($NV_{13}$ is closed) ($NV_{32}$ is closed) ($NV_{20}$ is medium) ($NV_2$ is open)
5. If ($V_1$ is closed) and ($V_{13}$ is closed) and ($V_{32}$ is closed) and ($V_{20}$ is closed) and ($V_2$ is closed) then ($NV_1$ is medium) ($NV_{13}$ is closed) ($NV_{32}$ is open) ($NV_{20}$ is medium) ($NV_2$ is closed)
6. If ($V_1$ is medium) and ($V_{13}$ is open) and ($V_{32}$ is closed) and ($V_{20}$ is medium) and ($V_2$ is closed) then ($NV_1$ is medium) ($NV_{13}$ is closed) ($NV_{32}$ is medium) ($NV_{20}$ is closed) ($NV_2$ is medium)

7. If ($V_1$ is medium) and ($V_{13}$ is closed) and ($V_{32}$ is medium) and ($V_{20}$ is open) and ($V_2$ is open) then ($NV_1$ is medium) ($NV_{13}$ is medium) ($NV_{32}$ is open) ($NV_{20}$ is open) ($NV_2$ is closed)
8. If ($V_1$ is closed) and ($V_{13}$ is medium) and ($V_{32}$ is open) and ($V_{20}$ is closed) and ($V_2$ is open) then ($NV_1$ is closed) ($NV_{13}$ is open) ($NV_{32}$ is medium) ($NV_{20}$ is closed) ($NV_2$ is closed)
9. If ($V_1$ is medium) and ($V_{13}$ is medium) and ($V_{32}$ is medium) and ($V_{20}$ is medium) and ($V_2$ is medium) then ($NV_1$ is closed) ($NV_{13}$ is medium) ($NV_{32}$ is medium) ($NV_{20}$ is closed) ($NV_2$ is open)
10. If ($V_1$ is medium) and ($V_{13}$ is medium) and ($V_{32}$ is closed) and ($V_{20}$ is closed) and ($V_2$ is closed) then ($NV_1$ is closed) ($NV_{13}$ is medium) ($NV_{32}$ is open) ($NV_{20}$ is closed) ($NV_2$ is closed)

For example, rule 1 states that if valve 1 and valve 13 and valve 32 are slightly open, valve 32 is half open, and the valve 20 and valve 2 are also slightly open, then the new values for the outputs of all the valves are that they will be slightly open, and other rules can be interpreted in a similar way.

The simulation was performed using different types of membership functions. Triangular, Gaussian and bell membership function s were used in the aggregator fuzzy system and the results are shown in Tables 4.2, 4.3 and 4.4. The error of each table is calculated using Eq. 4.1 but with different types of membership function.

**Table 4.2** Results using the triangular membership functions in the aggregator fuzzy system

| Valve | Error using triangular membership function |
|---|---|
| Valve 1 | 0.5369 |
| Valve 13 | 0.5369 |
| Valve 2 | 0.5369 |
| Valve 20 | 1.4631 |
| Valve 32 | 1.4631 |

**Table 4.3** Results using the Gaussian membership functions in the aggregator fuzzy system

| Valve | Error using Gaussian membership function |
|---|---|
| Valve 1 | 0.4421 |
| Valve 13 | 0.4421 |
| Valve 2 | 0.4421 |
| Valve 20 | 1.5579 |
| Valve 32 | 1.5579 |

**Table 4.4** Results using the bell membership functions in the aggregator fuzzy system

| Valve | Error using bell membership function |
|---|---|
| Valve 1 | 0.4739 |
| Valve 13 | 0.3848 |
| Valve 2 | 0.4532 |
| Valve 20 | 1.5038 |
| Valve 32 | 1.5368 |

Having this results we perform 30 evolution using the individual controllers type-1 fuzzy systems and results are shown in Table 4.5, in this table we illustrate 30 evolutions for each valve using Eq. 4.1 this is to observe the behavior of each valve.

We decide to optimize the individual controller using a genetic algorithm and results are illustrated in Table 4.6.

With the previous results we decided to optimized the parameters of the membership functions in the aggregator (inputs and outputs) and results are illustrated in Table 4.7.

**Table 4.5** Results using the type-1 individual fuzzy controllers

| Valve 1 | Valve 13 | Valve 2 | Valve 20 | Valve 32 |
|---|---|---|---|---|
| 0.7864 | 0.4743 | 0.5239 | 0.5215 | 0.6343 |
| 0.5959 | 0.6128 | 0.7136 | 0.4718 | 0.7965 |
| 0.8694 | 0.5102 | 0.645 | 0.5892 | 0.6486 |
| 1.0447 | 0.9318 | 0.823 | 0.823 | 0.7976 |
| 0.6887 | 0.8085 | 0.5598 | 0.6192 | 0.4842 |
| 0.5573 | 0.8079 | 0.5575 | 0.7933 | 0.4118 |
| 0.6667 | 0.4615 | 0.7084 | 0.5142 | 0.5311 |
| 0.6428 | 0.8186 | 0.9575 | 0.3879 | 0.3393 |
| 0.8247 | 0.6718 | 1.0955 | 0.8056 | 0.4376 |
| 1.0525 | 0.8530 | 1.098 | 0.5476 | 0.6292 |
| 0.8042 | 0.8979 | 1.0844 | 0.5684 | 0.3255 |
| 0.4494 | 0.9520 | 1.0289 | 0.7521 | 0.7745 |
| 0.7780 | 0.6679 | 0.671 | 0.6341 | 0.2257 |
| 0.4021 | 0.6618 | 0.7973 | 0.7774 | 0.6009 |
| 0.6879 | 0.7449 | 0.9034 | 0.462 | 0.6124 |
| 0.6432 | 0.7069 | 0.7258 | 1.096 | 0.5172 |
| 0.7590 | 0.6879 | 0.7044 | 0.8517 | 0.5216 |
| 1.0404 | 0.8798 | 0.8758 | 0.7772 | 0.554 |
| 0.8343 | 0.6014 | 0.7078 | 0.6553 | 0.9855 |
| 0.9223 | 0.6721 | 0.7381 | 0.5878 | 0.6048 |
| 0.6786 | 0.5999 | 0.5936 | 0.5943 | 0.3973 |
| 0.4929 | 0.8258 | 0.4361 | 0.6293 | 0.4468 |
| 0.8291 | 0.6187 | 0.6133 | 0.4118 | 0.6703 |
| 0.7849 | 0.6596 | 0.6789 | 0.6309 | 0.6776 |
| 0.7407 | 0.5966 | 0.5705 | 0.6035 | 0.5415 |
| 0.8589 | 0.9600 | 0.5934 | 0.8404 | 0.5572 |
| 0.8999 | 0.6375 | 0.6852 | 0.6966 | 0.5304 |
| 0.8747 | 0.5267 | 0.4607 | 0.4643 | 0.3716 |
| 0.6249 | 0.6306 | 0.6153 | 0.433 | 0.4844 |
| 0.8096 | 0.6108 | 0.6969 | 0.4499 | 0.9614 |

**Table 4.6** Results using the type-1 individual fuzzy controllers and genetic algorithm

| Valve 1 | Valve 13 | Valve 2 | Valve 20 | Valve 32 |
|---------|----------|---------|----------|----------|
| 0.1146 | 0.109 | 0.2077 | 0.0939 | 0.218 |
| 0.1228 | 0.131 | 0.1861 | 0.1329 | 0 |
| 0.1275 | 0.119 | 0.239 | 0.111 | 0 |
| 0.1116 | 0.115 | 0.2216 | 0.1092 | 0 |
| 0.0908 | 0.109 | 0.214 | 0.1191 | 0 |
| 0.1132 | 0.109 | 0.1922 | 0.0954 | 0 |
| 0.1225 | 0.117 | 0.1853 | 0.1003 | 0 |
| 0.1102 | 0.107 | 0.1938 | 0.1146 | 0 |
| 0.0993 | 0.105 | 0.2428 | 0.0851 | 0 |
| 0.1196 | 0.125 | 0.1433 | 0.113 | 0 |
| 0.1191 | 0.123 | 0.246 | 0.1394 | 0 |
| 0.1114 | 0.115 | 0.1539 | 0.091 | 0 |
| 0.1231 | 0.117 | 0.1818 | 0.101 | 0 |
| 0.1444 | 0.107 | 0.1366 | 0.0661 | 0 |
| 0.1225 | 0.117 | 0.1853 | 0.1003 | 0 |
| 0.1165 | 0.129 | 0.1883 | 0.1151 | 0 |
| 0.1225 | 0.117 | 0.1853 | 0.1003 | 0 |
| 0.1881 | 0.117 | 0.2333 | 0.1199 | 0 |
| 0.1225 | 0.117 | 0.1853 | 0.1003 | 0 |
| 0.1165 | 0.129 | 0.1883 | 0.1151 | 0 |
| 0.1165 | 0.129 | 0.1883 | 0.1151 | 0 |
| 0.1225 | 0.117 | 0.1853 | 0.1003 | 0 |
| 0.1225 | 0.117 | 0.1853 | 0.1003 | 0 |
| 0.1225 | 0.117 | 0.1853 | 0.1003 | 0 |
| 0.1165 | 0.129 | 0.1883 | 0.1151 | 0 |
| 0.133 | 0.129 | 0.2337 | 0.15 | 0 |
| 0.1296 | 0.123 | 0.2678 | 0.1078 | 0 |
| 0.1225 | 0.117 | 0.1853 | 0.1003 | 0 |
| 0.1165 | 0.129 | 0.1883 | 0.1151 | 0 |
| 0.1282 | 0.121 | 0.2337 | 0.1414 | 0 |

The last tables show the behavior of each valve using the proposed method and using the optimization, and then we decided to test the aggregator using a type-2 fuzzy system and the optimization to observe the behavior and compared the results. The behavior using the type-2 fuzzy aggregator is presented in Table 4.8.

**Table 4.7** Results using the type-1 fuzzy aggregator and genetic algorithm

| Valve 1 | Valve 13 | Valve 2 | Valve 20 | Valve 32 |
|---------|----------|---------|----------|----------|
| 0.0175  | 0.0176   | 0.0175  | 0.115    | 0.0727   |
| 0.017   | 0.0172   | 0.0171  | 0.12     | 0.075    |
| 0.9856  | 0.9831   | 0.993   | 0.993    | 0.065    |
| 0.0175  | 0.0176   | 0.017   | 0.1      | 0.0727   |
| 0.0955  | 0.09     | 0.093   | 0.098    | 0.0695   |
| 0.0176  | 0.0176   | 0.0176  | 0.1148   | 0.073    |
| 0.0176  | 0.0177   | 0.0172  | 0.115    | 0.0728   |
| 0.0176  | 0.0177   | 0.0173  | 0.1148   | 0.073    |
| 0.0175  | 0.0175   | 0.0174  | 0.115    | 0.0728   |
| 0.0177  | 0.0176   | 0.0172  | 0.1149   | 0.0732   |
| 0.0175  | 0.0175   | 0.0172  | 0.1148   | 0.0728   |
| 0.0175  | 0.0175   | 0.0172  | 0.1152   | 0.0731   |
| 0.0176  | 0.0176   | 0.0172  | 0.1149   | 0.0733   |
| 0.0176  | 0.0177   | 0.0172  | 0.1151   | 0.0725   |
| 0.0176  | 0.0176   | 0.0174  | 0.1148   | 0.0731   |
| 0.0175  | 0.0176   | 0.0173  | 0.115    | 0.073    |
| 0.0176  | 0.0176   | 0.0175  | 0.1149   | 0.0728   |
| 0.0176  | 0.0176   | 0.0173  | 0.1148   | 0.0729   |
| 0.0176  | 0.0175   | 0.0172  | 0.115    | 0.073    |
| 0.0176  | 0.0176   | 0.0172  | 0.1148   | 0.0728   |
| 0.0176  | 0.01746  | 0.0173  | 0.1148   | 0.0731   |
| 0.0176  | 0.0176   | 0.0175  | 0.1149   | 0.0728   |
| 0.0176  | 0.0176   | 0.0173  | 0.1148   | 0.0729   |
| 0.0175  | 0.0176   | 0.0173  | 0.1148   | 0.0731   |
| 0.0176  | 0.0175   | 0.0172  | 0.115    | 0.0729   |
| 0.0176  | 0.0177   | 0.0173  | 0.1152   | 0.0727   |
| 0.0175  | 0.0176   | 0.0173  | 0.1148   | 0.0732   |
| 0.0176  | 0.0176   | 0.0174  | 0.1148   | 0.073    |
| 0.0176  | 0.0176   | 0.0173  | 0.1148   | 0.0728   |
| 0.0176  | 0.0176   | 0.0175  | 0.1149   | 0.0726   |

Having the previous results a statistical comparison was performed to compare the behaviors, the results used the t student test using type-1 fuzzy system without optimization and type-1 fuzzy system using genetic algorithm for the valves and are shown in Tables 4.9, 4.10, 4.11, 4.12 and 4.13.

**Table 4.8** Error using triangular membership functions and genetic algorithm in type-2 fuzzy system

| Valve 1 | Valve 13 | Valve 2 | Valve 20 | Valve 32 |
|---|---|---|---|---|
| 0.0029 | 0.0028 | 0.0151 | 0.0205 | 0.0111 |
| 0.0028 | 0.0028 | 0.0152 | 0.0205 | 0.0111 |
| 0.003 | 0.0029 | 0.0152 | 0.0206 | 0.0115 |
| 0.0028 | 0.0028 | 0.015 | 0.0205 | 0.0115 |
| 0.0028 | 0.0028 | 0.0154 | 0.0205 | 0.0112 |
| 0.0029 | 0.0028 | 0.0152 | 0.0208 | 0.0113 |
| 0.0028 | 0.0028 | 0.0156 | 0.0205 | 0.0111 |
| 0.0029 | 0.0028 | 0.0158 | 0.0204 | 0.0112 |
| 0.0028 | 0.0028 | 0.0153 | 0.0205 | 0.0112 |
| 0.0029 | 0.0028 | 0.0157 | 0.024 | 0.0112 |
| 0.0028 | 0.0028 | 0.0153 | 0.0205 | 0.0112 |
| 0.003 | 0.0029 | 0.0152 | 0.0205 | 0.0111 |
| 0.0029 | 0.0028 | 0.0152 | 0.0208 | 0.0113 |
| 0.0028 | 0.0031 | 0.0155 | 0.0204 | 0.0113 |
| 0.003 | 0.0028 | 0.0156 | 0.0208 | 0.0112 |
| 0.0028 | 0.0028 | 0.0151 | 0.0204 | 0.0111 |
| 0.0029 | 0.0028 | 0.0157 | 0.0204 | 0.0112 |
| 0.0028 | 0.0028 | 0.0153 | 0.0205 | 0.0111 |
| 0.003 | 0.0028 | 0.0154 | 0.0204 | 0.0113 |
| 0.0028 | 0.0028 | 0.0152 | 0.0208 | 0.0111 |
| 0.0029 | 0.0031 | 0.0154 | 0.0208 | 0.0112 |
| 0.0029 | 0.0028 | 0.0151 | 0.0205 | 0.0111 |
| 0.0029 | 0.0029 | 0.0152 | 0.0205 | 0.0111 |
| 0.0028 | 0.0029 | 0.0159 | 0.0205 | 0.0111 |
| 0.003 | 0.0028 | 0.0156 | 0.0204 | 0.0111 |
| 0.0029 | 0.0028 | 0.0152 | 0.0208 | 0.0113 |
| 0.0028 | 0.0028 | 0.0154 | 0.0205 | 0.0112 |
| 0.0028 | 0.0028 | 0.015 | 0.0205 | 0.0115 |
| 0.003 | 0.0029 | 0.0152 | 0.0206 | 0.0115 |
| 0.0029 | 0.0028 | 0.0151 | 0.0205 | 0.0111 |

### 4.1.1 Statistical Comparison Using Type-1 Fuzzy System Without Optimization and Type-1 Fuzzy System Using Genetic Algorithm

Individual valve plot of valve 1 with the type-1 fuzzy systems and the type-1 fuzzy system with genetic algorithm is shown in Fig. 4.13. The value "T" is a distribution where hypothesis is null or not, "P" value is the probability of obtaining a result at

**Table 4.9** Results for the t student test using a type-1 fuzzy system without optimization and a type-1 fuzzy system using genetic algorithm in valve 1

| Fuzzy system | N | Mean | Deviation std. | Mean of std. error |
|---|---|---|---|---|
| Type-1 | 30 | 0.755 | 0.164 | 0.030 |
| Type-1 and GA | 30 | 0.1216 | 0.0158 | 0.0029 |

T = 21.03, P = 0.000, GL = 29

**Table 4.10** Results for the t student test using type-1 fuzzy system without optimization and type-1 fuzzy system using genetic algorithm in valve 13

| Fuzzy system | N | Mean | Deviation std. | Mean of std. error |
|---|---|---|---|---|
| Type-1 | 30 | 0.703 | 0.141 | 0.026 |
| Type-1 and GA | 30 | 0.11873 | 0.00759 | 0.0014 |

T = 22.68, P = 0.000, GL = 29

**Table 4.11** Results for the t student test using type-1 fuzzy system without optimization and type-1 fuzzy system using genetic algorithm in valve 2

| Fuzzy system | N | Mean | Deviation std. | Mean of std. error |
|---|---|---|---|---|
| Type-1 | 30 | 0.729 | 0.183 | 0.033 |
| Type-1and GA | 30 | 0.1984 | 0.0302 | 0.0055 |

T = 15.69, P = 0.000, GL = 30

**Table 4.12** Results for the t student test using type-1 fuzzy system without optimization and type-1 fuzzy system using genetic algorithm in valve 20

| Fuzzy system | N | Mean | Deviation std. | Mean of std. error |
|---|---|---|---|---|
| Type-1 | 30 | 0.633 | 0.163 | 0.030 |
| Type-1 and GA | 30 | 0.1090 | 0.0171 | 0.0031 |

T = 17.54, P = 0.000, GL = 29

**Table 4.13** Results for the t student test using type-1 fuzzy system without optimization and type-1 fuzzy system using genetic algorithm in valve 32

| Fuzzy System | N | Mean | Deviation std. | Mean of std. error |
|---|---|---|---|---|
| Type-1 | 30 | 0.569 | 0.175 | 0.032 |
| Type-1 and GA | 30 | 0.0073 | 0.0398 | 0.0073 |

T = 17.10, P = 0.000, GL = 31

least as extreme as the one that really was obtained and GL is the degrees of freedom.

Individual value plot of valve 13 with the type-1 fuzzy system and the type-1 fuzzy system with genetic algorithm is shown in Fig. 4.14.

Individual value plot of valve 2 with the type-1 fuzzy system and the type-1 fuzzy system with genetic algorithm is shown in Fig. 4.15.

Individual value plot of valve 20 with the type-1 fuzzy system and the type-1 fuzzy system with genetic algorithm is shown in Fig. 4.16.

Individual value plot of valve 32 with the type-1 fuzzy system and the type-1 fuzzy system with genetic algorithm is shown in Fig. 4.17.

**Fig. 4.13** Individual value
plot of valve 1

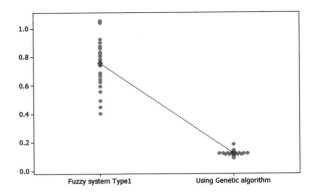

**Fig. 4.14** Individual value
plot of valve 13

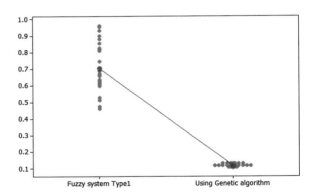

**Fig. 4.15** Individual value
plot of valve 2

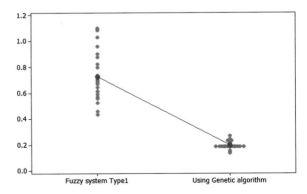

**Fig. 4.16** Individual value
plot of valve 20

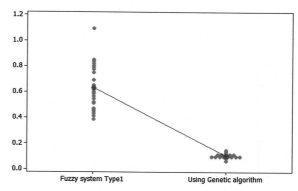

**Fig. 4.17** Individual value
plot of valve 32

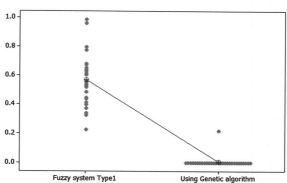

## 4.1.2   Statistical Comparison Using the Empirical Type-2 Fuzzy System and a Type-2 Fuzzy System Using Genetic Algorithm

It is important to consider a comparison between the empirical type-2 fuzzy system and the type-2 fuzzy system designed using a genetic algorithm, because is necessary to know if a significant difference exist when a genetic algorithm is applied to the type-2 fuzzy system design. Results for the t student test using type-2 fuzzy system and type-2 fuzzy system using genetic algorithm are presented in Tables 4.14, 4.15, 4.16, 4.17 and 4.18.

**Table 4.14** Results for the t student test using type-2 fuzzy system without optimization and type-2 fuzzy system using genetic algorithm in valve 1

| Fuzzy system | N | Mean | Deviation std. | Mean of std. error |
|---|---|---|---|---|
| Type-2 | 30 | 0.01639 | 0.00182 | 0.00033 |
| Type-2 and GA | 30 | 0.0028767 | 0.0000774 | 0.000014 |

T = 40.61, P = 0.000, GL = 29

**Table 4.15** Results for the t student test using type-2 fuzzy system without optimization and type-2 fuzzy system using genetic algorithm in valve 13

| Fuzzy system | N | Mean | Deviation std. | Mean of std. error |
|---|---|---|---|---|
| Type-2 | 30 | 0.01754 | 0.00139 | 0.00025 |
| Type-2 and GA | 30 | 0.0028367 | 0.0000809 | 0.000015 |

T = 57.82, P = 0.000, GL = 29

**Table 4.16** Results for the t student test using type-2 fuzzy system without optimization and type-2 fuzzy system using genetic algorithm in valve 2

| Fuzzy system | N | Mean | Deviation std. | Mean of std. error |
|---|---|---|---|---|
| Type-2 | 30 | 0.01755 | 0.00139 | 0.00025 |
| Type-2 and GA | 30 | 0.015337 | 0.0000240 | 0.000044 |

T = 8.56, P = 0.000, GL = 30

**Table 4.17** Results for the t student test using type-2 fuzzy system without optimization and type-2 fuzzy system using genetic algorithm in valve 20

| Fuzzy system | N | Mean | Deviation std. | Mean of std. error |
|---|---|---|---|---|
| Type-2 | 30 | 0.011696 | 0.00929 | 0.0017 |
| Type-2 and GA | 30 | 0.020663 | 0.000645 | 0.00012 |

T = 56.63, P = 0.000, GL = 29

**Table 4.18** Results for the t student test using type-2 fuzzy system without optimization and type-2 fuzzy system using genetic algorithm in valve 32

| Fuzzy system | N | Mean | Deviation std. | Mean of std. error |
|---|---|---|---|---|
| Type-2 | 30 | 0.2076 | 0.0233 | 0.0042 |
| Type-2 And GA | 30 | 0.011217 | 0.000134 | 0.000024 |

T = 46.22, P = 0.000, GL = 29

Individual value plot of valve 1 with the type-2 fuzzy system and the type-2 fuzzy system with genetic algorithm is shown in Fig. 4.18.

Individual value plot of valve 13 with the type-2 fuzzy system and the type-2 fuzzy system with genetic algorithm is shown in Fig. 4.19.

Individual value plot of valve 2 with type-2 fuzzy system and type-2 fuzzy system designed with genetic algorithm is shown in Fig. 4.20.

Individual value plot of valve 2 with the type-2 fuzzy system and the type-2 fuzzy system designed with genetic algorithm is shown in Fig. 4.21.

Individual value plot of valve 2 with type-2 fuzzy system and type-2 fuzzy system designed with genetic algorithm is shown in Fig. 4.22.

**Fig. 4.18** Individual value plot of valve 1

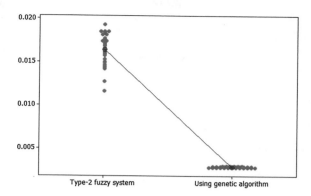

**Fig. 4.19** Individual value plot of valve 13

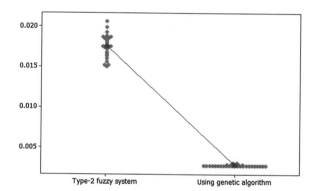

**Fig. 4.20** Individual value plot of valve 2

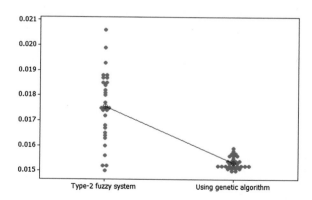

**Fig. 4.21** Individual value
plot of valve 20

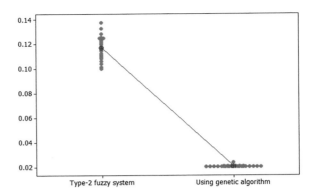

**Fig. 4.22** Individual value
plot of valve 32

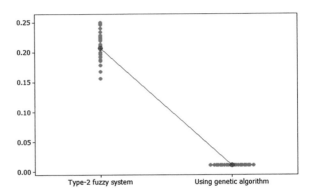

### 4.1.3   Statistical Comparison Using Type-1 Fuzzy System and Type-2 Fuzzy System with Genetic Algorithms

When genetic algorithms are used many times, better results are obtained, in this case of study the genetic algorithm helps to optimize the membership functions of the type-1 and type-2 fuzzy systems and obtain a better control in all the valves. First, the genetic algorithm was implemented in the 5 type-1 fuzzy systems. Then the genetic algorithm was used to optimize the type-2 fuzzy system. After obtaining 30 experiments using the genetic algorithm the t student tests were performed and the results are shown in Tables 4.19, 4.20, 4.21, 4.22, 4.23 and 4.24. The behavior

**Table 4.19** Results for the t student test using type-1 fuzzy system and type-2 fuzzy system using a genetic algorithm in valve 1

| Fuzzy System | N | Mean | Deviation std. | Mean of the std. error |
|---|---|---|---|---|
| Type-1 with GA | 30 | 0.1216 | 0.0158 | 0.0029 |
| Type-2 with GA | 30 | 0.0028767 | 0.0000774 | 0.000014 |

T = 41.10, P = 0.000, GL = 29

**Table 4.20**  Results for the t student test using type-1 fuzzy system and type-2 fuzzy system using a genetic algorithm in valve 13

| Fuzzy system | N | Mean | Deviation std. | Mean of std error |
|---|---|---|---|---|
| Type-1 with GA | 30 | 0.11873 | 0.00759 | 0.0014 |
| Type-2 with GA | 30 | 0.0028367 | 0.0000809 | 0.00015 |

T = 83.65, P = 0.000, GL = 29

**Table 4.21**  Results for the t student test using type-1 fuzzy system and type-2 fuzzy system using a genetic algorithm in valve 2

| Fuzzy system | N | Mean | Deviation std. | Mean of std. error |
|---|---|---|---|---|
| Type-1 with GA | 30 | 0.1984 | 0.0302 | 0.0055 |
| Type-2 with GA | 30 | 0.015337 | 0.000024 | 0.000044 |

T = 33.17, P = 0.000, GL = 29

**Table 4.22**  Results for the t student test using type-1 fuzzy system and type-2 fuzzy system using a genetic algorithm in valve 20

| Fuzzy system | N | Mean | Deviation std. | Mean of std. error |
|---|---|---|---|---|
| Type-1 with GA | 30 | 0.109 | 0.0171 | 0.0031 |
| Type-2 with GA | 30 | 0.020663 | 0.000645 | 0.00012 |

T = 22.22, P = 0.029, GL = 44

**Table 4.23**  Results for the t student test using type-1 fuzzy system and type-2 fuzzy system using a genetic algorithm in valve 32

| Fuzzy system | N | Mean | Deviation std. | Mean of std. error |
|---|---|---|---|---|
| Type-1 with GA | 30 | 0.0073 | 0.0398 | 0.0073 |
| Type-2 with GA | 30 | 0.011217 | 0.000134 | 0.000024 |

T = −0.54, P = 0.591, GL = 29

**Table 4.24**  Results of the simulation plant using triangular membership function

| | Elevator | Aileron | Rudder |
|---|---|---|---|
| Error | 1.9721E−35 | 1.9611E−035 | 4.44E−20 |

of each valve can change and in some valves the error decreases in type-1 more than type-2 that is because type-2 fuzzy system can model in a better way the uncertainty in complex problem with noise and in this case the problem is without noise.

The individual value plot of valve 1 with the type-1 fuzzy system and the type-2 fuzzy system designed is shown in Fig. 4.23.

The individual value plot of valve 13 with the type-1 fuzzy system and the designed type-2 fuzzy system is shown in Fig. 4.24.

**Fig. 4.23** Individual value
plot of valve 1

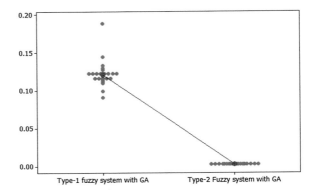

**Fig. 4.24** Individual value
plot of valve 13

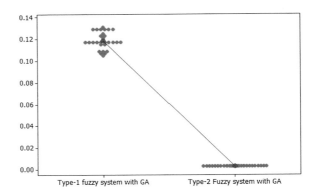

The individual value plot of valve 2 with the type-1 fuzzy system and the type-2
fuzzy system is shown in Fig. 4.25.

Individual value plot of valve 20 with the type-1 fuzzy system and the type-2
fuzzy system is shown in Fig. 4.26.

**Fig. 4.25** Individual value
plot of valve 2

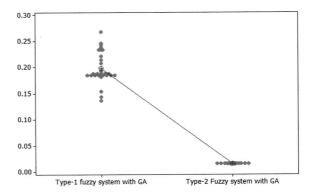

**Fig. 4.26** Individual value plot of valve 20

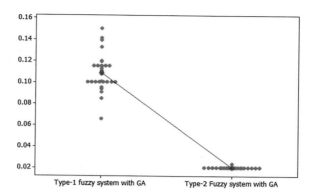

**Fig. 4.27** Individual value plot of valve 32

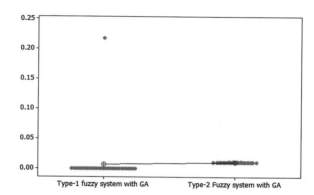

Individual value plot of valve 32 with the type-1 fuzzy system and the type-2 fuzzy system is shown in Fig. 4.27.

Result in tables (Sect. 4.1) shown that, there is statistical evidence to say that there is a significant difference in the results presented with regards to granular type-2 (more than 95 % confidence). In other words, the type-2 fuzzy system with granular computing generated a significant increase in the controller to have better control.

## 4.2 Simulation Results in the Second Case of Study Using F-16 Airplane

In this problem we need to control the total flight and for this reason is necessary to know that to control the flight is necessary to have 3 controllers (longitudinal, lateral and direction control) [1, 2]. For longitudinal control is important to consider the stick of the pilot because with this the elevator are moved (up and down pulling or pushing the stick) and this movement changes the position of the nose of the airplane (movement in axes *y*).

To perform direction control, the pedals make changes and when the pedals are pressed (left or right) the rudder moves (left or right) and the aircraft is moving left or right in axes *x*. Finally, the lateral control changes by moving the stick, left or right, when this happens the ailerons are moved up or down (alternating) and the aircraft moves in the axes *z*. Based on the explanation above, we decided to develop first the individual controllers (Longitudinal, lateral and direction control). A fuzzy system was design for each controller. To obtain the longitudinal control, a fuzzy system with 1 input and 1 output was designed, in the output variable is the elevator because when it moves the aircraft moves in the y position and in the input the Stick is used. In this case the linguistic variables are "push", "pull" or "not moved" the stick and depending on the movement a reaction in the elevator is obtained. In the lateral control the outputs are the ailerons because when the ailerons are moved, the aircraft can change the direction in the axes *z* and in the input are the stick of the pilot and when this is moved (right or left) the ailerons have a reaction (up/down) and the linguistic variables are the stick (move right, move left, center). In direction control the output is the rudder when the rudder is move (left or right) the aircraft change the trajectory in axes *x*, in the input the pedals are consider (left and right). When pedals are pressed the rudder has a reaction to left or right and the aircraft changes the position to the new direction. The fuzzy systems used for the 3 controllers are shown in Figs. 4.28, 4.29 and 4.30.

In this case of study we test using 2 different simulation plants to prove the method, the first simulation plant used is illustrated in Fig. 4.31.

The original simulation plant in the block of the pilot has a PID controller that provides the control of the elevator, ailerons and the rudder to maintain the flight

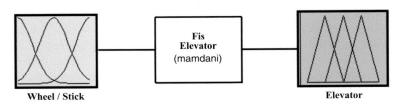

**Wheel / Stick**                                                      **Elevator**

**Fig. 4.28** Fuzzy system to longitudinal control

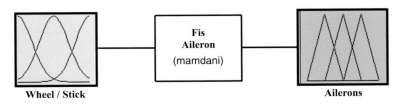

**Wheel / Stick**                                                      **Ailerons**

**Fig. 4.29** Fuzzy system to lateral control

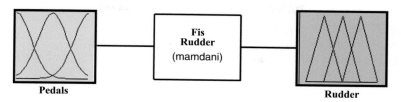

**Fig. 4.30** Fuzzy system to direction control

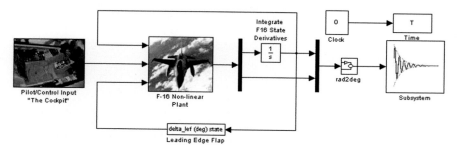

**Fig. 4.31** First simulation plant

**Fig. 4.32** Internal structure of the pilot control

control. In this work the block of the pilot was changed for the fuzzy systems and a part of the PID to achieve the control. The internal structure of the pilot control of the simulation plant is shown in Fig. 4.32.

Having the simulation plant and the fuzzy system, the simulation was performed using triangular membership functions in all the inputs and the outputs of the fuzzy system.

## 4.2.1   Simulation Results in the Second Case of Study Using F-16 Airplane

First the simulation was performed with a 20,000 ft value for the altitude, 650 ft/s in velocity, 10 deg for the elevator disturbance deflection, in aileron disturbance deflection of 10 deg and in rudder disturbance deflection of 10 deg. And the condition is steady wings-level flight. The behavior of the elevator, aileron and the rudder are shown in Figs. 4.33, 4.34 and 4.35, the yellow line is the reference and the purple line is the behavior.

The errors of the control are shown in Table 4.25.

The behavior in the flight control when the fuzzy system is applied is shown in Figs. 4.36, 4.37, 4.38 and 4.39.

**Fig. 4.33** Behavior of the elevators

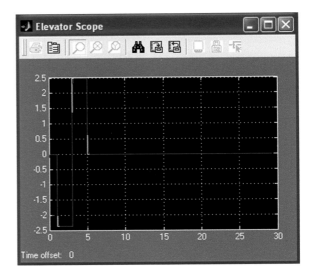

**Fig. 4.34** Behavior of the ailerons

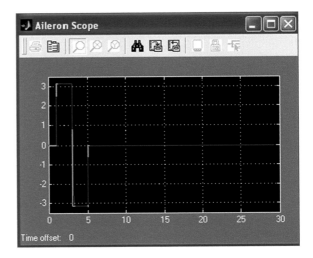

**Fig. 4.35** Behavior of the rudder

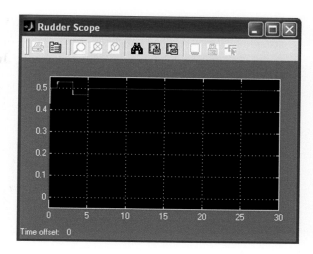

**Table 4.25** Results of the simulation plant using triangular membership function

|        | Elevator  | Aileron  | Rudder   |
|--------|-----------|----------|----------|
| Error  | 8.88E−20  | 2.22E−20 | 2.11E−20 |

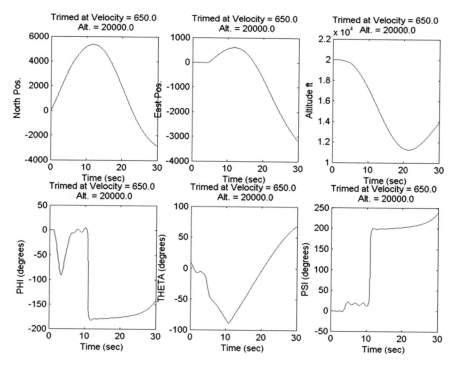

**Fig. 4.36** Behavior of flight control

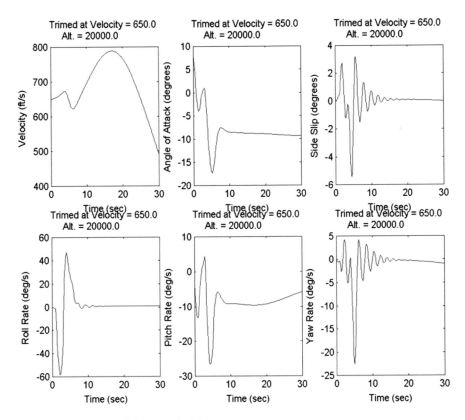

**Fig. 4.37** Behavior in flight control of the 3 axes

The previous figures show the changes during the 30 s, changes in the pitch, roll, yaw, and the behavior in the north and east position.

After having results with the simulation plant with 20,000 ft in altitude and 650 ft/s in velocity, then the altitude was increased from 20,000 to 40,000 ft and the velocity was increased from 650 to 900 ft/s, this is to observe the behavior of the airplane when the velocity and the altitude are increased and to observe if the fuzzy system can maintain the stability of the airplane while it is on flight.

In this case a $-15°$ in elevator disturbance deflection was used to observe the behavior. The behavior of the elevator, aileron and the rudder are shown in

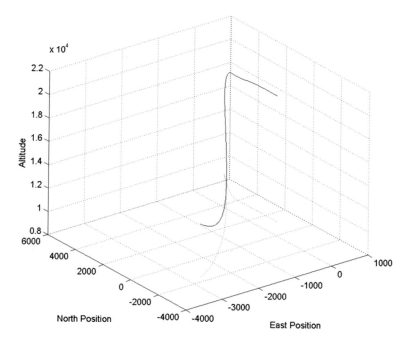

**Fig. 4.38** Behavior in flight in north and east position

Figs. 4.40, 4.41 and 4.42. The yellow lines is the reference and the purple lines are the behavior.

When the altitude and the velocity are modified the behavior of the airplane is different because it needs to modify the angles of the elevator, aileron and the rudder to have a stable flight. The results for this case are shown in Table 4.26.

During the flight the angles and the behavior of the elevator, aileron and rudder can change and these changes and movement are shown in Figs. 4.43, 4.44, 4.45 and 4.46.

This last figures presented the simulation with 40,000 ft in altitude and changes in the velocity and in the perturbations in the angle is noteworthy the behavior in the positions in the 30 s and the difference when simulation used 2000 ft in altitude is noticeable, and also the behavior of the pitch, roll and yaw are different. In Fig. 4.45 the movement of the airplane is presented, it shows the behavior in the north position, and the changes in the east position depending on the time of

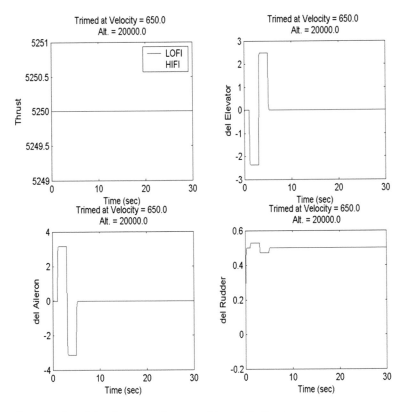

**Fig. 4.39** Behavior of the elevators, ailerons, rudder and in thrust

**Fig. 4.40** Behavior of the elevators

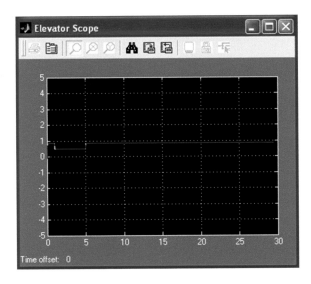

**Fig. 4.41** Behavior of the ailerons

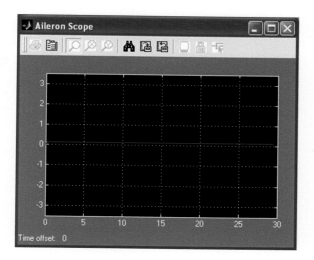

**Fig. 4.42** Behavior of the rudder

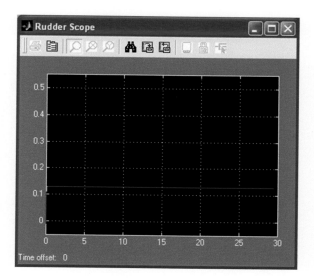

simulation and the altitude in this case the altitude started in 40,000 ft, and in the Fig. 4.45 the behavior is shown during the 30 s that drive the simulation.

We studied more thoroughly the above simulation plant and we observed that these had parameters very well established and for this reason we chose to use a second simulation plant that could difficult more the control flight of an aircraft to test the proposed method.

**Table 4.26** Results for the
simulation plant with
individual controller

| Ailerons | Elevator | Rudder |
|----------|----------|--------|
| 0.9142 | 1.0481 | 0.8881 |
| 0.9457 | 1.1064 | 0.932 |
| 0.8901 | 1.1133 | 0.9515 |
| 0.9121 | 1.0851 | 0.9446 |
| 0.8918 | 1.117 | 0.9098 |
| 0.8598 | 1.0906 | 0.8914 |
| 0.9441 | 1.0878 | 0.9258 |
| 0.9592 | 1.1512 | 0.8962 |
| 0.9045 | 1.1158 | 0.9305 |
| 0.9033 | 1.0465 | 0.9521 |
| 0.9032 | 1.0300 | 0.9230 |
| 0.8798 | 1.1215 | 0.9089 |
| 0.9200 | 1.1415 | 0.8965 |
| 0.8698 | 1.1536 | 0.9436 |
| 0.8544 | 1.0765 | 0.8836 |

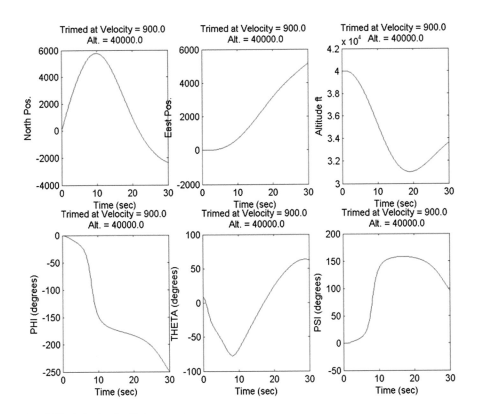

**Fig. 4.43** Behavior in flight control

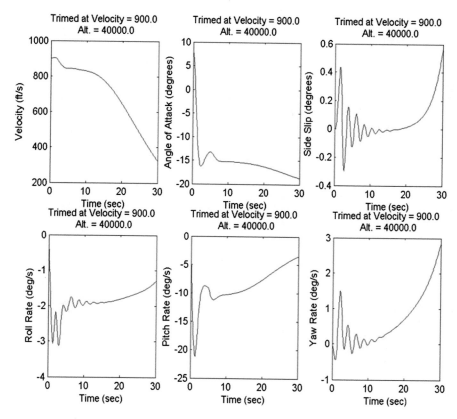

**Fig. 4.44** Behavior in flight control of the 3 axes

**Fig. 4.45** Behavior in flight in north and east position

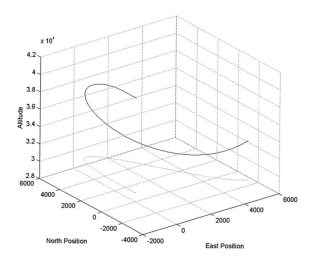

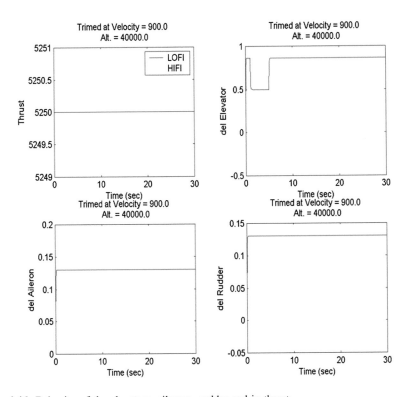

**Fig. 4.46** Behavior of the elevators, ailerons, rudder and in thrust

## 4.2.2  Simulation Results in Flight Control Using a DeHavilland Beaver

The second simulation plant used for this case of study is shown in Fig. 4.47.

And rules of each fuzzy system was defined based on knowledge about the flight and are shown in Figs. 4.48, 4.49 and 4.50.

Is important to mention that the simulation plant has a reference that has to follow the aircraft and a turbulence block but also has a joystick and this introduce more turbulence in the behavior of the aircraft in flight. For a better explanation the structure of the problem is shown in Fig. 4.51.

The results using the individual controllers are shown in Table 4.26.

The behavior of the controllers is shown in Fig. 4.52.

The last figure shows the behavior of the elevator, aileron and rudder, this results were obtained using manually executions to observe the behavior later we show the behavior using a genetic algorithm and its improvement. In Table 4.26 can appreciate that errors are high, then in this part of the book we show how the

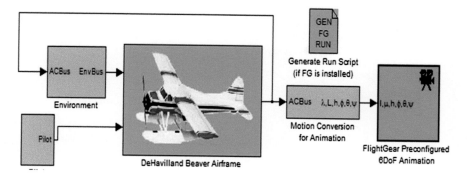

**Fig. 4.47**  Fuzzy system to direction control

> + If wheel is "push" then elevator is down
> + If wheel is center then elevator is center
> + If wheel is "pull" then elevator is up

**Fig. 4.48**  Rules for the longitudinal control

> + If wheel is left then aileron is left_up_right_down
> + If wheel is center then aileron is center
> + If  wheel is right then aileron is right_up_left_down

**Fig. 4.49**  Rules for the lateral control

> + If pedals is left then rudder is left
> + If pedals is center then rudder is center
> + If pedals l is right then rudder is right

**Fig. 4.50**  Rules for the direction control

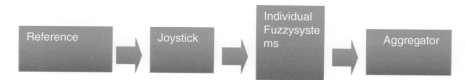

**Fig. 4.51**  Structure of the problem

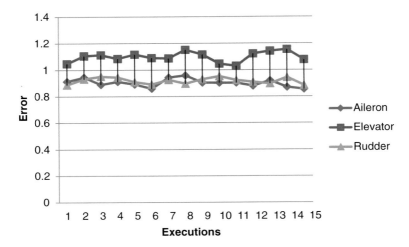

**Fig. 4.52** Behavior of the airplane

proposed method is apply. First remembering that the 3 controllers was obtain, each controller has one output. Then we develop an aggregator with 3 inputs and three outputs, the inputs take the values of each individual controller to obtain new outputs to achieve the control. The aggregator is shown in Fig. 4.53.

Having the aggregator the rules set is needed to continue and how we can know what are the best rules?, for this reason a genetic algorithm [3, 4] was applied to obtain the best rules set, and the chromosome of the genetic algorithm is shown in Fig. 4.54.

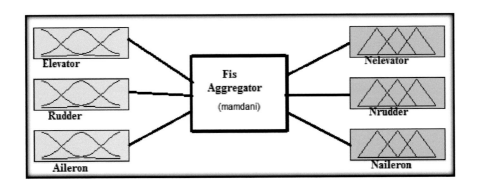

**Fig. 4.53** Aggregator of the proposed method

| Antecedent | | | Consequent | | |
|---|---|---|---|---|---|
| 1 | 2 | 3 | 4 | 5 | 6 |

**Fig. 4.54** Chromosome of the genetic algorithm for rules set

To obtain the rules we use the Michigan approach where each chromosome is a rule. Each gene contains a value between 1 and 3, and the parameters are: number of generations—20, number of individuals (population size) 25, selection-roulette, discrete Mutation, crossover of a single point with a crossover rate = 0.9 and mutation rate—0.2. Finally the rules set are as follows:

1. If (Elevator is *high*) and (rudder is *high*) and (Aileron is *low*) then (Nelevator is *high*) (Nrudder is *medium*) (Naileron is *medium*)
2. If (Elevator is *high*) and (rudder is *medium*) and (Aileron is *high*) then (Nelevator is *medium*) (Nrudder is *high*) (Naileron is *medium*)
3. If (Elevator is *low*) and (rudder is *medium*) and (Aileron is *low*) then (Nelevator is *low*) (Nrudder is *low*) (Naileron is *medium*)
4. If (Elevator is *medium*) and (rudder is *high*) and (Aileron is *medium*) then (Nelevator is *medium*) (Nrudder is *medium*) (Naileron is *medium*)
5. If (Elevator is *medium*) and (rudder is *medium*) and (Aileron is *high*) then (Nelevator is *medium*) (Nrudder is *medium*) (Naileron is *medium*)
6. If (Elevator is *medium*) and (rudder is *low*) and (Aileron is *high*) then (Nelevator is *low*) (Nrudder is *medium*) (Naileron is *low*)
7. If (Elevator is *low*) and (rudder is *medium*) and (Aileron is *high*) then (Nelevator is *medium*) (Nrudder is *high*) (Naileron is *medium*)
8. If (Elevator is *low*) and (rudder is *low*) and (Aileron is *low*) then (Nelevator is *medium*) (Nrudder is *medium*) (Naileron is *high*)
9. If (Elevator is *low*) and (rudder is *high*) and (Aileron is *low*) then (Nelevator is *medium*) (Nrudder is *medium*) (Naileron is *medium*)
10. If (Elevator is *low*) and (rudder is *medium*) and (Aileron is *medium*) then (Nelevator is *medium*) (Nrudder is *medium*) (Naileron is *medium*)

Having the rules and the aggregator ready the simulation was perform and results are shown in Table 4.27.

The behavior of the airplane is illustrated in Fig. 4.55 where the reference and behavior are shown.

**Table 4.27** Results for the
simulation plant using the
aggregator

| Aileron | Elevator | Rudder |
| --- | --- | --- |
| 0.4185 | 0.3881 | 0.1877 |
| 0.5047 | 0.5044 | 0.3047 |
| 0.5031 | 0.5033 | 0.5020 |
| 0.4073 | 0.3611 | 0.1626 |
| 0.5052 | 0.5023 | 0.5049 |
| 0.3600 | 0.3740 | 0.2684 |
| 0.2468 | 0.2417 | 0.2964 |
| 0.2413 | 0.2654 | 0.4503 |
| 0.3595 | 0.3784 | 0.5396 |
| 0.5043 | 0.5041 | 0.5079 |
| 0.5039 | 0.5045 | 0.508 |
| 0.5042 | 0.5053 | 0.5046 |
| 0.3529 | 0.3785 | 0.3601 |
| 0.5071 | 0.5056 | 0.5121 |
| 0.2288 | 0.2430 | 0.5022 |

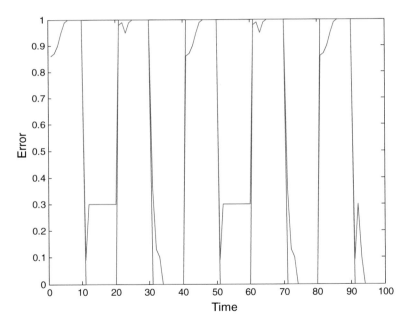

**Fig. 4.55** Behavior of the airplane

When the type-1 aggregator was adapt as a proposed method the behavior was better but we decide to use a type-2 aggregator to performed the behavior of the aircraft. Is important to mentioned that the aggregator is a type-2 fuzzy system to achieve the control and the system was optimize with genetic algorithm. Results applying the type-2 aggregator in the proposed method are presented in Table 4.28.

The performance of the genetic algorithm is presented in Fig. 4.56.

**Table 4.28** Results using the type-2 fuzzy aggregator

| Aileron | Elevator | Rudder |
|---|---|---|
| 0.20925 | 0.19405 | 0.09385 |
| 0.25235 | 0.2522 | 0.15235 |
| 0.25155 | 0.25165 | 0.251 |
| 0.20365 | 0.18055 | 0.0813 |
| 0.2526 | 0.25115 | 0.25245 |
| 0.18 | 0.187 | 0.1342 |
| 0.1234 | 0.12085 | 0.1482 |
| 0.12065 | 0.1327 | 0.22515 |
| 0.17975 | 0.1892 | 0.2698 |
| 0.25215 | 0.25205 | 0.25395 |
| 0.25195 | 0.25225 | 0.254 |
| 0.2521 | 0.25265 | 0.2523 |
| 0.17645 | 0.18925 | 0.18005 |
| 0.25355 | 0.2528 | 0.25605 |
| 0.1144 | 0.1215 | 0.2511 |

**Fig. 4.56** Performance of the genetic algorithm

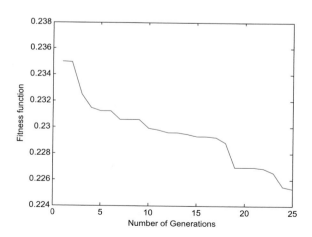

#### 4.2.2.1  Statistical Comparison Using the Individual Controllers and the Aggregator

Also the behavior of each controller when the proposed method was applied is better and the aggregator works better to improve the control, but to test the aggregator is useful to perform a statistical $t$ student, 6 test were made one for each controller (aileron, elevator and rudder). The first statistical test is using the individual value of the aileron with its 15 samples and the values obtained with the aggregator and genetic algorithm, the second test is with the individual values for elevator controller compared with the aggregator and the genetic algorithm, and finally the last test is with individual value of the rudder controller and the aggregator with optimization, then the last three tests are between type-1 aggregator fuzzy system and the type-2 fuzzy system.

In Table 4.3 the test shows that there is statistical evidence to say that there is significant difference in the results presented in this work with respect to the type-2 aggregator (with more than 95 % confidence). And the behavior is shown in Fig. 4.57 where values are plotted.

In Tables 4.29, 4.30 and 4.31 the improvement is shown when the proposed method is apply, in comparisons between the individual fuzzy system and the

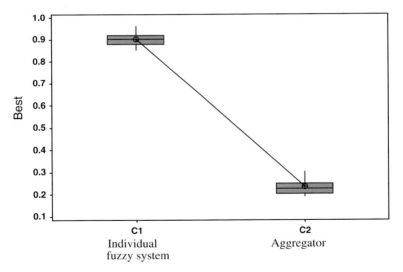

**Fig. 4.57** Box plot for individual values of ailerons and values of the type-2 aggregator of aileron

**Table 4.29** Results for the *t* student test simulation with the aileron

|  | N | Mean | Deviation std. | Mean of std. error |
|---|---|---|---|---|
| Individual fuzzy controller for aileron | 15 | 0.9035 | 0.0309 | 0.0080 |
| Aggregator | 15 | 0.2049 | 0.0533 | 0.014 |

*Results* T-value = 43.94, P-value = 0.000, DF = 28

**Table 4.30** Results for the *t* student test simulation with the elevator

|  | N | Mean | Deviation std. | Mean of std. error |
|---|---|---|---|---|
| Individual fuzzy controller for elevator | 15 | 1.0990 | 0.0377 | 0.0097 |
| **Aggregator** | 15 | 0.2053 | 0.0510 | 0.013 |

*Results* T-Value = 54.61 P-Value = 0.000 DF = 28

**Table 4.31** Results for the *t* student test simulation with rudder

|  | N | Mean | Deviation std. | Mean of std. error |
|---|---|---|---|---|
| Individual fuzzy controller for rudder | 15 | 0.9185 | 0.0238 | 0.0062 |
| Aggregator | 15 | 0.2037 | 0.0655 | 0.017 |

*Results* T-value = 39.73, P-value = 0.000, DF = 28

type-2 aggregator with optimization, and that significant differences were also observed with a margin of more than 95 % confidence.

#### 4.2.2.2 Statistical Comparison with the Type-1 and Type-2 Aggregator

Also a statistical t student was performed in comparison with type-1 and type-2 aggregator and results are shown in Tables 4.32, 4.33 and 4.34.

The previous results represented the behavior of different control problems when the proposed method is applied (Fig. 4.58).

**Table 4.32** Results for the *t* student test simulation with the aileron for type-1 and type-2

|  | N | Mean | Deviation std. | Mean of std. error |
|---|---|---|---|---|
| Type-1 fuzzy system aggregator | 15 | 0.410 | 0.107 | 0.028 |
| Type-2 fuzzy system aggregator | 15 | 0.2049 | 0.0533 | 0.014 |

*Results* T-value = 6.66, P-value = 0.000, DF = 28

**Table 4.33** Results for the *t* student test simulation with the elevator for type-1 and type-2

|                               | N  | Mean   | Deviation std. | Mean of std. error |
|-------------------------------|----|--------|----------------|--------------------|
| Type-1 fuzzy system aggregator | 15 | 0.410  | 0.102          | 0.026              |
| Type-2 fuzzy system aggregator | 15 | 0.2053 | 0.0510         | 0.013              |

*Results* T-value = 6.98, P-value = 0.000, DF = 28

**Table 4.34** Results for the *t* student test simulation with the rudder for type-1 and type-2

|                               | N  | Mean   | Deviation std. | Mean of std. error |
|-------------------------------|----|--------|----------------|--------------------|
| Type-1 fuzzy system aggregator | 15 | 0.407  | 0.131          | 0.034              |
| Type-2 fuzzy system aggregator | 15 | 0.2037 | 0.0655         | 0.017              |

*Results* T-value = 5.39, P-value = 0.000, DF = 28

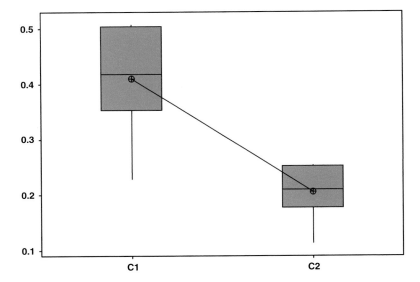

**Fig. 4.58** Box plot for individual values of the ailerons and values of the type-2 and type-1 aggregator of the aileron

In tables, the improvement is shown when the proposed method is apply, in other words, the proposed method generated a significant increase in the controller's performance to have better stability against perturbations. In the comparisons between the individual fuzzy system and the aggregator with optimization is illustrated in tables, and that significant differences were also observed with a margin of more than 95 % confidence.

# References

1. Dadios, E.: Fuzzy Logic-Controls, Concepts, Theories and Applications (2012)
2. Haupt, R., Haupt, S.: Practical Genetic Algorithm. Wiley Interscience, Hoboken (2004)
3. Holland, J.H.: Adaptation in Natural and Artificial Systems. University of Michigan Press, Ann Arbor (1975)
4. Oh, S.-K., Jung, S.-H., Pedrycz, W.: Design of optimized fuzzy cascade controllers by means of hierarchical fair competition-based genetic algorithms. Expert Syst. Appl. **36**(9), 11641–11651 (2009)

# Chapter 5
# Conclusions

In this book the main objective was to develop a method to achieve the behavior in complex control problems, as a first point we designed the architecture to the proposed method. This architecture had to be useful to implement in any control problem with more than 2 controllers.

Having designed the architecture of the proposed method, different cases of control were studied and we selected as the first case of control to prove the proposed method. The first case was the water control using three tanks and five valves connected between the tanks. To know if the proposed method could give good results, first the control problem was studied with more detail. These details were obtained working first with individual controllers using type-1 and type-2 fuzzy systems and optimization with genetic algorithms.

Then the proposed method was applied using type-1 and type-2 fuzzy systems, in this case when this architecture was used in the problem of the three tanks, when the proposed method was implemented, the results obtained were better than when the problem is studied individually or as one controller. Is important to mention that genetic algorithms were implemented in the architecture to achieve the parameters and obtain a better control.

When we obtained results using the three tanks water control we decided to prove the method using a more complex control problem. The second case of study was the flight of an airplane. In this case of study we had many types of airplanes to choose, first we selected one airplane F-16, this simulation plant was studied and we observed that the simulation plant have not many complexity to control it, and we decided to use a different simulation plant but using the flight control.

In the second case of study using the second simulation plant (flight control), we decided to introduce disturbances (turbulences) using a generator block in Matlab. This block has equations that generate the turbulences when the airplane is in flight. Also we decided to introduce more noise using a physical joystick to prove the proposed method using a complex control problem.

Having the complex control problem (second case of study) we applied the same methodology used with the first case of study. First we obtain results of the control problem individually and we observed the behavior of the flight using type-1 and type-2 fuzzy systems and also genetic algorithms, then the proposed method was applied and results were better than without the proposed architecture.

L. Cervantes and O. Castillo, *Hierarchical Type-2 Fuzzy Aggregation of Fuzzy Controllers*, SpringerBriefs in Computational Intelligence, DOI 10.1007/978-3-319-26671-8_5

Is important to mention that the proposed method was a good alternative in complex control problems (with more than one individual controller) and in this case this architecture was proved with two cases of control and we need to know if results were good. Having results in both cases we decided to use a statistical test to prove the results. In both cases results show that when the proposed method is applied, the error decreases and with this we can say that the proposed method is a good alternative to achieve the control and obtain better results.

As future work the proposed method can be tested in real cases using physical elements such as a physical airplane, also the parameters of the proposed method can be achieved using a different kind of optimization, and obtain the optimization of each parameter in the method.

We recommend using more optimization methods to achieve the architecture of the proposed method. Based on the results obtained we recommend to use this proposed method applying in different fields of study in our case we used fuzzy logic control.

# Appendix

When the second case of study was applied we used an interface using Matlab with a flight simulator to observe the behavior in a real time using different sceneries, some of this are illustrated in next figures. The last 3 figures show an example of successful landing, the last figure illustrate the view from the cockpit in good landing. Some of the figures show sceneries with clouds, day or night and also buildings.

© The Author(s) 2016
L. Cervantes and O. Castillo, *Hierarchical Type-2 Fuzzy Aggregation of Fuzzy Controllers*, SpringerBriefs in Computational Intelligence,
DOI 10.1007/978-3-319-26671-8

# Index

© The Author(s) 2016
L. Cervantes and O. Castillo, *Hierarchical Type-2 Fuzzy Aggregation
of Fuzzy Controllers*, SpringerBriefs in Computational Intelligence,
DOI 10.1007/978-3-319-26671-8

Printed in the United States
By Bookmasters